WORLD INDUSTRY STUDIES 6

The World Mining Industry

Investment Strategy and Public Policy

WORLD INDUSTRY STUDIES

Edited by Professor Ingo Walter,
Graduate School of Business Administration,
New York University

The World Mining Industry

Investment Strategy and Public Policy

Raymond F. Mikesell
University of Oregon

John W. Whitney
Whitney & Whitney Inc. (Consultancy), Reno, Nevada

Boston
ALLEN & UNWIN
London   Sydney   Wellington

338.2
M63w

© Raymond F. Mikesell and John W. Whitney, 1987
This book is copyright under the Berne Convention. No reproduction
without permission. All rights reserved.

Allen & Unwin, Inc.,
8 Winchester Place, Winchester, Mass. 01890, USA
the US company of
UNWIN HYMAN LTD
PO Box 18, Park Lane, Hemel Hempstead, Herts HP2 4TE, UK
40 Museum Street, London WC1A 1LU, UK
37/39 Queen Elizabeth Street, London SE1 2QB, UK

Allen & Unwin (Australia) Ltd,
8 Napier Street, North Sydney, NSW 2060, Australia

Allen & Unwin (New Zealand) Ltd in association with the Port
Nicholson Press Ltd,
60 Cambridge Terrace, Wellington, New Zealand

First published in 1987

Library of Congress Cataloging-in-Publication Data

Mikesell, Raymond Frech.
 The world mining industry.
(World industry studies; 6)
Bibliography: p.
Includes index.
 1. Mineral industries – Finance.   2. Mineral
 industries – Government policy.
I. Whitney, John W. (John Wallis)
II. Title.   III. Series.
HD9506.A2M525 1987   338.2   87–1763
ISBN 0-04-338120-0 (alk. paper)

British Library Cataloguing in Publication Data

Mikesell, Raymond F.
 The world mining industry: investment strategy and
public policy. – (World industry studies; 6)
1. Mineral industries
I. Title   II. Whitney, John W.   III. Series
338.2 HD9506.A2
ISBN 0-04-338120-0

Typeset in 10 on 11 point Times by Oxford Print Associates Ltd.
and printed in Great Britain by
Billing & Sons Ltd, London and Worcester

This book is dedicated to our wives,
Irene and Reenie

University Libraries
Carnegie Mellon University
Pittsburgh, Pennsylvania 15213

Contents

List of Tables

List of Figures

Foreword

Few industries have experienced such economic and political shocks as the global mining sector. In the mid-1980s much of the industry is depressed. Large new ore bodies discovered and developed a decade earlier have hit their stride, while structural change involving lighter and smaller manufactured goods and the growth of service industries has made economic growth much less material-intensive than in earlier times. The resulting supply glut for many materials is a far cry from the widespread shortages predicted by the Club of Rome just a few years ago. Prices are depressed and industries, as well as countries, are facing hard times.

None of this makes world mining any less interesting, or any less critical for the future of the international economy. Supply and demand relationships will surely change again, resurrecting some of the concerns that plagued policymakers a decade ago. And the industry remains a fascinating amalgam of special problems. Modern mining projects tend to be exceedingly large in scale. They require complex and risky forms of financing. They face all kinds of difficult environmental and logistical problems. Their impact on local and national economies can be as immense as it is complex. As a consequence, they tend to be politically sensitive, and negotiations between companies and governments are among the most interesting to be found in international business.

In this volume, Messrs Mikesell and Whitney have provided a concise, authoritative overview of this complex industry. Both have been heavily involved in the industry's problems for most of their working lives, and this is clearly reflected in the narrative. Their objective is to present the reader with a careful and thorough non-technical analysis, a clear picture of a unique industry of central importance to the international economy. In this the authors succeed admirably.

INGO WALTER
New York University

Preface

This book applies the principles of management and financial strategy to the mining industry. In addition, it provides an economic and technical analysis of the world mining industry and of the directions in which it is moving. This study combines the knowledge and experience of an academician who has specialized in the economics of the minerals industry and in international finance, and a mine management consultant and mineral economist with experience in the major mining countries of the Western world. The two authors have been collaborating on consulting assignments for mining firms and on publications on mining for over a decade.

The authors gratefully acknowledge the contribution of William A. Fuchs of Whitney & Whitney in the preparation of the material on exploration in Chapter 3, and that of James E. Zinser, Professor of Economics, Oberlin College for the preparation of the case study on Rosario Dominicana in Chapter 5. They also acknowledge the opportunity provided by Professor Ingo Walter to write this book and for his advice and assistance.

1

An Overview of the World Mining Industry

General Characteristics

The mining industry is defined in various ways, depending upon what materials are covered and the degree of processing of the materials extracted from the earth. According to the broadest definition, mining includes discovering, extracting and processing of all non-renewable resources up to the point at which they are used as inputs for fabricating or for producing energy. This broad definition includes the energy minerals such as coal, petroleum and natural gas; refined or processed metals such as copper, steel and the ferroalloys; and nonminerals such as diamonds, phosphate and potash. A much narrower definition of mining includes only crude or nonprocessed mine products, such as mineral ores and coal, and excludes petroleum and natural gas. In this study we deal mainly with the major metals from the exploration and mining stages to the processing stage from which they are normally marketed for use in manufacturing. The major metals include iron and the ferroalloys (nickel, manganese, chromium and molybdenum); and the nonferrous metals (aluminum, copper, lead, gold, silver, tin and zinc). However, we shall not deal with the iron and steel industry beyond the iron ore mining stage or with the aluminum industry beyond the bauxite mining stage since these industries would require large studies in themselves.

The production of metals involves several stages that are generally carried on by large mining firms, although small mining operations may engage in the initial stage. The first stage is exploration of areas identified by geological reports as possessing potential mineral resources. Modern exploration methods are quite sophisticated and include geological, geochemical and geophysical investigation; three-dimensional sampling by core drilling or other methods; laboratory analyses, including ore treatment, concentration, and recovery tests;

and economic appraisal. The objective is to discover and evaluate an orebody that can be economically exploited.

Geochemical exploration is used to measure the chemical properties of the area surrounding the deposit in order to delineate abnormal chemical patterns that may be related to potentially economic mineral deposits. Geophysical investigations employ electronic equipment that can detect subtle contrasts in such physical properties as specific gravity, electrical conductivity, heat conductivity, seismic velocity and magnetic susceptibility. Where much of the bedrock is concealed, telegeologic or remote sensing techniques measure various geologic properties from aircraft or satellites. Exploration is commonly carried on by teams of specialists that include geologists, geochemists and geophysicists. There are different levels of exploration beginning with regional geologic mapping of areas up to 50,000 square km (20,000 square miles) and ending with intensive investigations of orebodies by means of numerous drillings to obtain bulk samples which are then metallurgically tested to determine the dimensions and character of the orebody.

If the results of exploration activities suggest that an economical deposit has been found, the second stage is conducting a feasibility study which involves engineering and economic evaluations of the mining project. It is on the basis of this study that companies decide whether to go ahead with a mining project; the study may also be reviewed by prospective lenders. The feasibility study for a large mining project may be quite costly, running to $25 million or more in some cases. The total cost of exploration and the feasibility study for a large mine may run to $50 million or more. It is uncertain whether a profitable mine will be constructed until all the stages have been completed. In the initial exploration stage, several million dollars may be spent with less than a 10 per cent chance of a successful outcome.

The third stage is the construction of the mine, the metallurgical plant, and infrastructure. There are two basic types of operations to extract mineral ores: open-pit or surface mining, and underground mining. An open-pit mine is largely a quarrying operation that handles a large volume of material. Such mining involves drilling and blasting the ore and hauling it out of the pit in large trucks with capacities ranging up to 200 tons, or in ore trains. The ore is hauled to crushers and then to the metallurgical plant. In underground mining, shafts are dug into ore deposits below the surface, from which ore is drilled, blasted and removed through underground passages to the surface. Iron, bauxite and copper ores are extracted by means of open-pit mining, while lead, zinc, silver and gold are largely extracted by underground mining. There are also some underground copper mines. Economies of scale in open-pit mining permit the mining of relatively low-grade ores. As much as 100,000 tons of ore per day containing less than 1 per cent metal are extracted in the larger open-pit operations. Higher ore grades are necessary for underground mining to be profitable.

Large mines involve huge capital outlays running to a billion dollars or more. The mining complexes usually include beneficiation or concentration of ores for production of concentrates with 25 per cent or higher metal content. In the case of copper, large mine complexes include plants for smelting copper or for producing copper metal by hydrometallurgical methods, but in the case of other metals such as gold, lead, zinc, tin and iron, metal is produced in separate plants which may or may not be owned by the mining company. The degree of processing that usually takes place at the mine differs widely among metals, but refining the product for marketing to fabricators nearly always takes place in separate plants that refine the products of several mines.

Since mines tend to be located in mountains, deserts, or jungles and away from developed areas, infrastructure is often a substantial proportion of capital cost. It is frequently necessary to provide sources of power and water, as well as highways, railroads and port facilities. In addition the mining company may be responsible for constructing living quarters for workers and their families and for providing education and other public services required by the mining community. In developing countries especially, it is usually necessary to provide a training program for the nationals who work in the mine.

The Importance of World Production of Minerals

Table 1.1 shows 1985 world mine production (metal content) of twenty-two nonfuel minerals together with the *reserve base* (measured reserves that can be economically produced plus marginal reserves and some subeconomic resources that may be economically produced in the future).[1] The relative importance of each of these minerals in the world economy must be determined by the value of annual production based on market prices. The reserve base provides a rough indication of the number of years a mineral can be produced at current levels of output and in most cases the reserve base is sufficient to support current annual production for a number of decades. While consumption is increasing for each of these minerals, the reserve base is constantly being expanded by the discovery of new resources. For most minerals the rate of increase in the reserve base has been greater than the rate of increase in annual production.

Table 1.1   *World Mine Production and Reserve Base, 1985[a]*

	Production		Reserve base	
Bauxite	76,300	mt	23,200,000	mt
Chromite	10,600	st	7,500,000	st
Cobalt	35,100	st	9,200,000	st
Columbium	29,000	thousand lb	9,100,000	thousand lb
Copper	7,805	mt	525,000	mt
Diamonds (industrial)	37.8	million carats	990	million carats
Gold	47.0	million troy oz	1,450	million troy oz
Ilmenite	4,615	thousand st	734,000	thousand st
Iron ore	799	million st	98,000	million st
Lead	3,350	mt	143,000	mt
Manganese	25,800	st	12,000,000	st
Mercury	188,000	76-lb flasks	7,200,000	76-lb flasks
Molybdenum	209,500	thousand lb	25,950,000	thousand lb
Nickel	821,000	st	111,000,000	st
Platinum group	7,400	thousand troy oz	1,200,000	thousand troy oz
Rutile	392	thousand st	133,400	thousand st
Silver	394	million troy oz	10,800	million troy oz
Tantalum	710	thousand lb	76,000	thousand lb
Tin	201,000	mt	3,000,000	mt
Tungsten	45,100	mt	3,460,000	mt
Vanadium	71,000	thousand lb	36,500,000	thousand lb
Zinc	6,560	thousand mt	300,000	thousand mt

[a] estimated

Note: mt = metric ton; st = short ton

Source: Bureau of Mines, *Mineral Commodity Summaries 1986*, Washington, DC: US Department of Interior, 1986.

The total value of world crude mineral production in 1978 has been estimated by F. G. Callot at about $479 billion, of which 67 per cent represented petroleum and natural gas; 18 per cent coal and lignite; 10 per cent metals; and 4 per cent nonmetals.[2] The 1978 value of mine production of forty-five major nonfuel minerals was estimated at $62 billion, of which 46 per cent was produced in developed countries; 25 per cent in developing countries; and 29 per cent in communist countries (mainly the Soviet Bloc and China).

Except for a few developing countries whose export incomes are mainly derived from nonfuel mineral production, mine production of nonfuel minerals represents a very small percentage of GNP. For the developed countries, the mine value of nonfuel minerals produced represented only 0.4 per cent of the aggregate of these countries' GNP, and for the developing countries about 0.8 per cent of GNP (Callot, 1981, p. 27).[3] The value of mine production of metals in the

USA in 1980 (based on the recoverable metal content of the ores) was \$8.9 billion, or 0.3 per cent of US GNP. However, these low percentages of GNP do not indicate the importance of nonfuel minerals in the world's economy since without them industrial production would soon come to a halt.

Table 1.2 shows the 1978 value of world mine production of twenty of the most important nonfuel minerals ranked by value of mine production. The mine value of the twenty minerals was \$5.5 billion, or about 90 per cent of the total value of world mine production of nonfuel minerals. Between 1978 and 1984 world mine production of metals declined by about 3 per cent and the combined real price index of the ten most important minerals declined by about 9 per cent. Therefore, the 1984 value of world mine production of the twenty nonfuel minerals was probably less than \$50 billion in 1978 prices.

Table 1.2 *World Production of Major Crude Minerals in 1978* (billions of dollars)

Mineral	Value	Ranking
Iron	\$11.6	1
Copper	8.6	2
Gold	7.5	3
Phosphates	3.0	4
Tin	2.5	5
Potash	2.5	6
Diamonds	2.0	7
Lead	2.0	8
Zinc	1.0	9
Asbestos	1.9	10
Silver	1.8	11
Bauxite	1.6	12
Nickel	1.5	13
Sulphur	1.3	14
Platinum	1.2	15
Molybdenum	1.0	16
Manganese	0.9	17
Kaolin	0.8	18
Tungsten	0.8	19
Chromite	0.6	20
Others	7.0	
Total	\$62.0	

Source: F. G. Callot, "World Mineral Production and Consumption in 1978", *Resources Policy*, March 1981, p. 16.

*Distribution of the Value of Nonfuel Minerals Production
by Major Country and Region in 1978*

Table 1.3 shows the ranking by value output and percentage of total value output of nonfuel minerals for twenty countries that accounted for virtually all nonfuel mine production in 1978. It will be observed that the leading producer is the USSR, which produced 32 per cent more crude nonfuel minerals by value than the USA, ranked second; South Africa, Canada and Australia ranked third, fourth and fifth respectively, and the combined value of mine output of nonfuel minerals of these three countries exceeded that of the USSR. If comparable data were available for 1984, they would undoubtedly show that both the USSR and the three countries named above increased their value output relative to that of the USA. Only two Western European countries, France and West Germany, are listed in Table 1.3 and both rank below Poland. It is somewhat surprising to note that developing countries together (excluding China) produced only 13.7 per cent of the total value output of crude nonfuel minerals

Table 1.3 *Major World Producers of Nonfuel Minerals Ranked by Value of Mine Output, 1978* (billions of dollars)

Country	Value	%	Ranking
USSR	$12.9	20.7	1
USA	8.8	14.1	2
South Africa	6.8	11.0	3
Canada	4.4	7.2	4
Australia	3.1	4.9	5
China	2.6	4.1	6
Chile	1.5	2.4	7
Brazil	1.4	2.2	8
Peru	1.0	1.6	9
India	1.0	1.5	10
Mexico	0.9	1.5	11
Zaire	0.9	1.4	12
Poland	0.9	1.4	13
France	0.8	1.3	14
Zambia	0.8	1.2	15
Malaysia	0.7	1.2	16
Morocco	0.7	1.1	17
West Germany (FRG)	0.6	1.0	18
Philippines	0.5	0.8	19
Japan	0.5	0.8	20
Others	0.1		

Source: F. G. Callot, "World Mineral Production and Consumption in 1978", *Resources Policy*, March 1981, p. 26.

as contrasted with 40.3 per cent for developed countries (including South Africa), and 26.2 per cent for the communist countries (including China).[4]

Table 1.4 shows the leading world producers of twenty-two major nonfuel minerals in 1985. It will be observed that the USA was among the five largest producers of eight of these minerals (copper, gold, iron ore, lead, mercury, molybdenum, silver and zinc), while the USSR was among the five largest producers of fifteen of these minerals. Moreover, the USSR ranked first or second in ten of these minerals, while the USA was first or second in only three (copper, lead and molybdenum). The importance of Australia, Canada and South Africa as leading producers of several of these commodities is worth noting. South Africa is among the first five leading producers for chromite, gold and manganese; Australia among the first five producers of bauxite, lead, zinc and nickel; while Canada is among the first five producers of copper, gold, lead, nickel and zinc.

The industrial countries of North America, Western Europe and Japan consume over 80 per cent of nonfuel minerals produced in the non-communist world, while the largest non-communist producers of thirteen of the twenty-two minerals listed in Table 1.4 are in the developing countries, plus South Africa. In 1985 developed countries produced the largest share of mine output of titanium (ilmenite and rutile), lead, mercury, molybdenum, nickel and zinc, and a substantial (but not majority) portion of bauxite, copper, iron ore, silver, tantalum, tungsten and vanadium. Developed countries also produced the bulk of the *refined* copper, ferromanganese, steel and other processed minerals, but a large proportion of the mine raw materials for these products comes from developing countries plus South Africa. Developing countries plus South Africa hold larger shares of the reserves of most important nonfuel minerals than developed countries. Moreover, the developing countries' share of world resources from which additional reserves will eventually be established by increased exploration is substantially larger than that of the developed countries.

Production and reserves of several important minerals are heavily concentrated in South Africa and the Central African countries of Gabon, Zaire, Zambia and Zimbabwe. These minerals include chromium, cobalt, industrial diamonds, gold, manganese, the platinum group metals (platinum and palladium) and vanadium. Tin production is heavily concentrated in the Far Eastern countries of Malaysia, Indonesia and Thailand. Production of bauxite, copper, iron ore, lead, nickel and zinc are more widely distributed geographically, with substantial production in both developed and developing countries.

Table 1.4 *Leading World Producers of Selected Nonfuel Mine Products, 1985*

Product	*Major country producers*
Bauxite	Australia, 35%; Guinea, 16%; Brazil, 8%; Jamaica, 7%; USSR, 6%
Chromite	South Africa, 31%; USSR, 31%; Albania, 9%; Turkey, 7%; India, 5%; Zimbabwe, 5%
Cobalt	Zaire, 51%; Zambia, 14%; USSR, 9%; Canada, 6%; Cuba, 5%
Columbium	Brazil, 83%; Canada, 17%
Copper	Chile, 17%; USA, 13%; Canada, 9%; USSR, 8%; Zaire, 7%
Diamonds (indus.)	Zaire, 35%; Botswana, 18%; USSR, 17%; South Africa, 15%; Australia, 7%
Gold	South Africa, 47%; USSR, 19%; Canada, 6%; USA, 5%; Australia, 4%
Ilmenite	Australia, 19%; Canada, 18%; Norway, 15%; USSR, 11%; South Africa, 10%
Iron ore	USSR, 30%; Australia, 16%; USA, 12%; Canada, 12%; Brazil, 8%
Lead	Australia, 14%; USA, 12%; Canada, 8%; Mexico, 6%; Peru, 6%
Manganese	USSR, 43%; South Africa, 15%; Brazil, 9%; Gabon, 9%
Mercury	USSR, 34%; Spain, 23%; Algeria, 12%; USA, 8%
Molybdenum	USA, 51%; Chile, 17%; Canada, 7%; Peru, 5%; Mexico, 4%
Nickel	USSR, 24%; Canada, 24%; Australia, 10%; Indonesia, 9%; New Caledonia, 9%
Platinum group	USSR, 50%; South Africa, 43%; Canada, 5%
Rutile	Australia, 51%; Sierra Leone, 26%; South Africa, 16%
Silver	Mexico, 16%; Peru, 14%; USSR, 12%; USA, 11%; Canada, 10%
Tantalum	Thailand, 35%; Australia, 25%; Brazil, 25%
Tin	Malaysia, 20%; USSR, 17%; Thailand, 10%; Indonesia, 10%; Bolivia, 9%
Tungsten	China, 30%; USSR, 20%; South Korea, 6%; Australia, 4%
Vanadium	South Africa, 42%; USSR, 31%; China, 17%
Zinc	Canada, 18%; Australia, 11%; Peru, 9%; Mexico, 5%; USA, 4%

Source: Bureau of Mines, *Mineral Commodity Summaries 1986*, Washington, DC: US Department of Interior, 1986, p. 89.

Concentration of Ownership and Control

Ownership and control of world mining is heavily concentrated in a small number of multinational mining firms (most of which are privately owned) and in state mining enterprises (SMEs). There are thousands of small, privately owned mining firms in the developed countries and in some of the major Latin American mining countries, such as Brazil, Chile, Mexico and Peru. However, small mines produce less than 25 per cent of world output and their activities tend to be concentrated in gold, silver, diamonds and other precious stones, and in types of mining where economies of scale are less important. The vast bulk of the world's output of bauxite, copper, iron ore, manganese, molybdenum and nickel is produced by large-scale operations in mines costing over $50 million. A substantial amount of gold, silver and platinum is produced as byproducts of large-scale copper and copper-nickel mining. Small- and medium-sized mines account for about 75 per cent of the output of chromite and tungsten in the market economy countries, over half of the output of lead and zinc, and nearly 30 per cent of the tin production (Leaming, 1983, pp. 64–5).

Prior to the 1960s, the bulk of the mining capacities of developing countries were owned by multinational mining firms. The wave of expropriations in Africa and Latin America during the 1960s and 1970s brought a large portion of bauxite, copper, iron ore and tin producing capacity under control of SMEs. In some countries, such as Brazil, SMEs came to dominate the national mining industry without expropriation of mining firms; while in Mexico mixed private domestic and state ownership replaced foreign ownership. By the early 1980s SMEs controlled nearly 65 per cent of the copper mining capacity of developing countries; multinational mining firms controlled less than 25 per cent; and the remainder was under private domestic control. In iron ore, nearly three-fourths of the output of developing countries was in the hands of SMEs, with perhaps no more than 10 per cent under effective control by multinationals, and the remainder was controlled by private domestic enterprise. In the case of tin, about half of Third World output is controlled by SMEs and the remainder divided between private domestic enterprises and multinationals. However, the bulk of Third World output of bauxite and nickel mining is effectively controlled by multinational companies. SMEs in Brazil, Ghana, Guinea and Jamaica own substantial interests in bauxite mines in partnership with multinational aluminum companies, but the multinationals provide the management and purchase most of the output.

Despite the expropriation of mining properties in developing countries, large multinational mining firms continue to own and control a substantial share of metal producing capacity in the non-communist world. These multinational firms are relatively few in number and have their headquarters in the USA, Canada, Australia, South Africa, or one of the Western European countries. The major multinational firms listed in Appendix Table 1.1a produce metals in both developed and developing countries. Most of these companies produce more than one metal or nonmetal mineral; most are integrated through the stages of production from mining to refining the metal, while a few of them (such as German and Japanese companies) confine their operations largely to smelting and refining of metals. The most important US multinational mining firms in terms of metal production are AMAX, Anaconda (Arco), ASARCO (American Smelting and Refining Company), Alcoa, Newmont, Kennecott (Standard Oil of Ohio), US Steel, Phelps Dodge, Kaiser Aluminum & Chemical, Exxon and St Joe Minerals (Fluor Corporation). Three British Firms, Rio Tinto Zinc (RTZ), Consolidated Gold Fields and Selection Trust (subsidiary of British Petroleum), produce a wide range of metals in a number of developed and developing countries. RTZ controlled assets in excess of $9 billion in 1982 and its subsidiaries accounted for the second largest volume of copper production of any corporate group in the world and are among the five largest producers of bauxite, zinc and tin. There are several major Canadian multinationals, including INCO (the world's largest nickel producer) and Alcan (one of the world's largest aluminum producers). Two Australian multinationals, Broken Hill Pty and Conzinc Rio Tinto of Australia (CRA), a subsidiary of RTZ, have substantial mining operations in the Asia-Pacific region. Anglo-American of South Africa (AAC) produces a wide range of minerals with investments throughout the world.

The companies listed in Appendix Table 1.1a account for well over half the non-communist world output of gold, bauxite-alumina-aluminum, nickel, molybdenum, diamonds, lead and zinc; over 40 per cent of the copper output; and a substantial portion of the world's output of iron ore, manganese, tungsten, chromium, vanadium, platinum-palladium, tin and a number of other metals. Although these companies have a substantial share of the nonfuel mineral output of developing countries, the largest share of their mining assets are in developed countries, including the USA, Canada, South Africa, Australia and Western Europe. Many of these multinational firms also hold minority equity interests in joint ventures with SMEs in developing countries, and in many cases two or more multinationals will have an equity interest in the same operating company. Finally, many of these multinational firms are major producers of petroleum, coal and uranium, as well as nonfuel minerals.

Six large integrated aluminum companies produced about 60 per cent of the non-communist world's aluminum output in 1980 and several are integrated into fabrication. The six largest producers of nickel accounted for about 70 per cent of the non-communist world's output of refined nickel or ferronickel. The largest multinational producers of tin, lead and zinc produce about half the non-communist world's refined output of each of these metals; while the largest multinational firms producing copper accounted for over one-third of the non-communist world's production of refined copper.

Entry of US Petroleum Companies
The financial structure of the world mining industry has been affected by the entry of large petroleum (or other nonmining) firms into metal mining during the last decade. This entry has taken the form of mergers between older multinational mining firms and large US petroleum firms, and of the direct entry of petroleum firms into metal mining. Examples of the former include the merger of Anaconda with Arco; of Kennecott Copper with Standard Oil of Ohio; of Freeport Minerals with McMoRan; and of St Joe Minerals with Fluor Corporation. Exxon and Standard Oil of Indiana, through its subsidiary Amoco Minerals, have established new mining operations in the USA and abroad. For example, Exxon owns and operates the large Disputada copper mine in Chile, while Amoco Minerals is a major partner in the Ok Tedi gold/copper mine in Papua New Guinea (PNG).

These firms that now control a substantial portion of the world mining industry have large amounts of capital to invest in contrast to the traditional mining firms whose financial resources in recent years have been reduced by low profits. This is especially significant in view of the high capital costs for modern mines. On the other hand, petroleum and other large companies that have recently entered the mining business are unlikely to continue to invest in mining unless profits are comparable with those in other industries. The new entrants have less dedication to exploration and expansion of their share of world output of particular metals than the traditional mining firms that possess a highly specialized group of geologists and mine managers. Moreover, an important purpose of some of the mergers was to acquire the undeveloped orebodies of the traditional mining firms as a means of resource asset diversification, since it is cheaper and less risky to buy ore reserves than to find them. This could have an adverse effect on exploration both in the USA and abroad.

Trends in Production, Consumption and Real Prices of Major Metals

World production and consumption of metals are sensitive to changes in the rates of growth of real GNP in the industrial countries. The average annual increase in the United Nations (UN) Index of World Metal Production was 5.1 per cent for the 1953–63 period and 5.7 per cent for the 1963–73 period, but declined to 0.2 per cent for the 1973–80 period. For the market economies the rate of growth in the index of metal production declined from 4.5 per cent per annum in the 1953–63 period to 3.7 per cent in the 1963–73 period, and became negative in the 1973–80 period (see Table 1.5).

From 1945 to the early 1970s, world demand for metals grew rapidly, paralleling the rapid growth of real GNP in the industrial countries. However, the rate of real GNP growth in the industrial economies declined from an annual average of 5.1 per cent during the 1960–73 period to 2.5 per cent during the 1973–80 period, with a further decline during the world recession beginning in 1981.

The average annual rate of growth in consumption of eight major minerals (weighted by value of consumption) in the non-communist countries declined from 6.4 per cent in the 1961–70 period to 2.4 per cent in the 1970–80 period. The average annual rate of growth in consumption of copper declined from 3.9 per cent to 2.5 per cent;

Table 1.5   *Average Annual Rates of Growth in World Production of Metals (in percentages)*

	1953–63	1963–73	1973–80
World[a]	5.1*	5.7	0.2
All market economies	4.5	3.7	−0.4
Developed[b]	4.7	3.2	−1.1
Developing[c]	4.1	5.0	0.9
Centrally planned economies[d]	11.5	13.0	2.5

* = based on the period 1955–63

[a] Excludes China, Mongolia, Democratic People's Republic of Korea, Democratic Republic of Vietnam, and Albania.

[b] Includes northern North America, Europe (excluding Eastern Europe), Australia, Israel, Japan, New Zealand and South Africa.

[c] Includes Caribbean, Central and South America, Africa (except South Africa), Asian Middle East, and East and South-East Asia (except Israel and Japan).

[d] Includes Bulgaria, Czechoslovakia, East Germany (GDR), Hungary, Poland, Romania and USSR.

Note: Based on the UN Metals Production Index which includes iron ore and thirty nonferrous metals, including gold, bauxite, chromium, cobalt, lead, manganese, mercury, molybdenum, nickel, platinum, silver, tantalum, tin, tungsten, uranium and zinc.

Source: United Nations, *Monthly Bulletin of Statistics*, New York: United Nations, various issues.

Table 1.6 *Consumption Growth in Nonfuel Minerals in the Market Economies, 1961–70, 1970–80 and 1980–95 (Projected) (per cent per annum)*

	1961–70	1970–80	Projected[A] 1980–95	Projected[B] 1983–2000
Copper[a]	3.9	2.5	2.6	2.7
Tin[a]	0.4	−0.8	0.2	1.0
Nickel[a]	7.6	2.2	2.4	2.9
Lead[a]	4.3	2.1	3.2	1.8
Zinc[a]	4.5	1.3	3.1	2.0
Aluminum[a]	9.9	4.3	3.9	4.0
Iron ore	5.4	0.8	2.5	2.4
Manganese ore[b]	8.6	1.8	3.0	n.a.
Average[c]	6.4	2.4	2.9	n.a.

[A] World Bank
[B] US Bureau of Mines
[a] Refined metal includes secondary material
[b] Manganese content basis
[c] Weighted by the value of consumption in industrial and developing countries in 1980
n.a. = not available
Source: World Bank, 1983, p. 87; and Bureau of Mines, 1986b.

iron ore from 5.4 per cent to 0.8 per cent; and aluminum from 9.9 per cent to 4.3 per cent (see Table 1.6). The projected rates of growth in consumption of most of the eight minerals for 1980–95 shown in Table 1.6 are somewhat higher than the actual rates during the 1970–80 period, but are well below the actual rates of growth for the 1961–70 period. The consumption growth projections made by the World Bank and the US Bureau of Mines (BOM) shown in Table 1.6 differ as a consequence of the use of different projection periods and of different methodologies by the two agencies. Some investigators regard the consumption growth projections for at least some metals as too high. For example, a World Bank staff study projects the annual rate of growth in copper consumption between 1984 and 1995 at 1.3 per cent per year (Takeuchi, Strongman, Maeda (1986), p. 117).

The decline in rates of growth in world consumption of minerals has been due to a combination of lower rates of growth in GNP in the industrial countries and a shift in the composition of consumption in favor of services. The latter trend is likely to continue in industrial countries, but in developing countries the rates of growth in demand for commodities with a high metal content, e.g. automobiles and refrigerators, are likely to be relatively high. In addition, new materials technology has led to conservation in the use of certain metals and to the substitution of nonmetallic materials (e.g. optical fibers) for metals.

The structural decline in demand for minerals during the 1970s, together with the cyclical decline during the 1981–3 recession, created a condition of world overcapacity in metals in the late 1970s and early 1980s. This led to a sharp fall in real prices of metals. Actually, there has been a long-term downward trend in real prices of most metals since the mid-1960s. Figure 1.1 shows movements in the combined (weighted) index of prices of ten major metals and minerals – copper, tin, nickel, bauxite, aluminum, iron ore, manganese ore, lead, zinc and phosphate rock – for the 1950–85 period in both current and constant 1981 dollars. The average index of prices (in 1981 dollars) in 1980–85 was 66 per cent of the average for the 1950–2 period; 59 per cent of the average for 1964–6; and 50 per cent below the average for 1972–4 (World Bank, 1986, p. 11). Thus not only were average mineral prices (in constant dollars) in 1980–5 at their lowest level in more than thirty years, but the trend has been downward since 1964–6.

There is considerable evidence that the real cost of producing metals, including capital costs of creating new capacity, has risen relative to most metal prices, and that in 1982–5 the prices of major minerals did not cover the full economic costs of production for most mines. Over the longer run, full economic costs must be reflected in higher prices if the metals industry is to replace existing capital equipment and depleted reserves, and expand capacity to meet long-run growth in demand for the products. For these reasons real prices of metals are almost certain to rise above the levels of the early 1980s. When metal prices will rise depends in considerable measure on the elimination of world overcapacity in most metals. A World Bank staff study of January 1986 forecasts the combined (weighted) index of prices (in constant dollars) of ten major metals and minerals in 1995 at 82 (1979–82 = 100), a level slightly lower than the average index for 1982–5 (World Bank, 1986, p. 11). Although this may be too pessimistic, it implies that overcapacity in the metals industries will exist for at least another decade. How much prices will rise when consumer demand and capacity are again in equilibrium will depend upon the marginal economic costs of expanding capacity or the full economic costs of the last unit of additional capacity required to meet consumer demand. Such projections are difficult to make because both capital and operating costs have been changing rapidly in recent years. A 1983 study by the BOM estimates that a copper price of $1.25 to $1.50 (in 1981 dollars) would be necessary to cover full economic costs of producing copper at the 1981 world level. This is more than twice the average price in 1985 (Rosenkranz, Boyle and Porter, 1983, p. 28). Other investigators believe that sufficient additional capacity to meet copper demand in the 1990s will be

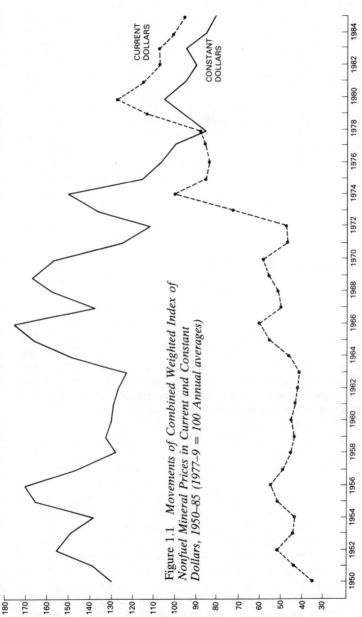

Figure 1.1 *Movements of Combined Weighted Index of Nonfuel Mineral Prices in Current and Constant Dollars, 1950–85 (1977–9 = 100 Annual averages)*

Note: The commodities in the index weighted by 1977–9 export values are: copper, tin, nickel, bauxite, aluminum, iron ore, manganese ore, lead, zinc and phosphate rock.

Source: World Bank, *Half-Yearly Revision of Commodity Price Forecasts and Quarterly Review of Commodity Markets for December 1985*, Washington, DC: World Bank, January 1986.

provided at a price of about $1 per pound in 1985 dollars (Crowson, 1983).

Although the real costs per pound of duplicating all the metal producing capacity in existence today may be twice the per pound cost of creating new capacity in the 1970s, a substantial portion of existing capacity will be producing metals well into the next century. The per unit cost of additional capacity required to meet the demand for metals during the remainder of this century will not be the average cost of duplicating existing capacity, but the per unit cost of expanding capacity by the lowest cost producers. Moreover, technological developments are continually taking place that reduce per unit costs of new capacity. Therefore, it is difficult to forecast the prices of metals required to induce sufficient capacity growth to meet future demands. Although it seems likely that real prices must rise above those in the first half of the 1980s, it appears unlikely that real prices will rise to the levels of the 1960s and early 1970s by the end of the present century.

Market Structure and Competition

Despite a high degree of concentration of production, world markets for minerals are generally quite competitive and most metal prices are subject to a high degree of fluctuation over the business cycle. Most major metals are traded on commodity exchanges – the London Metal Exchange (LME) or the New York Commodity Exchange (COMEX) or both. Although the volume of trade in metals on the LME and COMEX is relatively small compared to total trade in metals, quotations on these exchanges govern the contract prices between metal producers and buyers in domestic and international markets. In recent years growing competition from developing countries has reduced the ability of major mining and processing firms to maintain 'producer' prices in sales to their customers. There are also merchant markets in virtually all metals, which provide alternative sources of supply for buyers and outlets for small producers. In periods of declining demand and short supply, prices in the merchant markets may differ substantially from producer prices quoted by the other integrated suppliers of copper, nickel, zinc and other metals.

SMEs in developing countries tend to maintain capacity in periods of declining demand and sell all they can produce at world market prices, while producers in the USA, Canada and certain other developed countries have sought to maintain prices by reducing output in line with demand. For example, US copper producers reduced production by more than a third from 1981 to 1984 in response to the decline in the price of copper, while Chile's SME, Corporacion Nacional del Cobre (CODELCO), expanded copper output.

There are no commodity exchanges or well-organized markets for bauxite, iron ore, or for other ores and concentrates. Most of the output of iron ore and bauxite is both controlled and consumed by the integrated aluminum and steel companies. There is, however, a substantial amount of trade in copper, lead and zinc concentrates produced by companies without smelting or refining capacities. Concentrates are usually sold under long-term contracts at prices related to the world price of the metal less the processing costs, or the concentrates may be processed on a toll or fee basis, with ownership of the metal remaining with the seller.

The Adequacy of Mineral Reserves

Despite the doomsday forecasts of depletion of nonfuel mineral resources, or their scarcity in relation to growing demand during the present century, there appear to be ample resources for meeting the projected growth in demand well into the twenty-first century. Over the past half-century continuous exploration aided by technological advances has increased reserves of most nonfuel minerals more rapidly than they have been depleted, and much of the earth's surface has not been adequately explored by modern methods. The development of seabed minerals could conceivably provide ample supplies of copper, nickel, manganese, cobalt and other minerals for many generations to come.

Over the past two decades two developments have changed the outlook for the availability of nonfuel minerals. One has been the sharp decrease in the rate of growth in world demand for metals since the mid-1960s. The second has been the rapid advances in technology for (1) the production of minerals from lower ore grades without a significant rise in costs; and (2) the substitution of materials that are exceedingly abundant in the earth's crust, such as clays, for less abundant ones. Aluminum can be made from clays rather than bauxite; optic fibers made of sand can replace copper in communications; new ceramic materials made from clay have the potential to replace metals in engine blocks; and alloys made of nickel can replace cobalt in jet engines. There is a growing consensus that a scarcity of nonfuel minerals is unlikely to constitute a limitation on world economic growth, at least to 2030 (Manners, 1981, pp. 1–20). However, this optimistic assessment depends upon (1) the maintenance of relatively free markets for minerals; (2) the availability of mineral resources on land and under the ocean for exploration and development; and (3) a continuation of current rates of technological progress.

Import Dependence on Nonfuel Minerals

All nations are to some degree dependent on imports of nonfuel minerals. In the USA and Western Europe the degree of dependence on imports has increased with the exhaustion of higher grade ores of traditional industrial metals, such as copper, iron ore, lead, zinc and tin, and with the introduction of new metals, such as aluminum, chromium, cobalt, columbium, tantalum, titanium and vanadium, into the industrial process during the twentieth century. Import dependence increases not only with shifts in demand relative to domestic sources of supply, but with the discovery and development of cheaper sources in other countries. Thus while the USA is potentially self-sufficient in copper and iron ore, a portion of its requirements of these commodities is imported from foreign sources. In the early 1920s the USA was self-sufficient in bauxite for producing aluminum, but as the aluminum industry expanded and cheaper sources of bauxite were developed in other countries, the USA became dependent on imports for the vast bulk of its bauxite requirements.

Figure 1.2 shows US net import reliance on selected minerals and metals as a percentage of consumption in 1981, while Figure 1.3 shows net import reliance for the European Economic Community (EEC), Japan and the USSR in 1980. According to these charts, the USSR is significantly reliant on imports for only a few minerals, but unlike the market economy countries, it tends to strive for self-sufficiency with little regard for the relationship between world prices and domestic production costs.

Recycling and the use of scrap materials as inputs for the production of metals tend to reduce import dependence. Secondary recovery constitutes an important source of supply for certain metals. This source is limited by the cost of collecting and processing scrap, and for most metals this source cannot be counted on to provide more than 10 to 15 per cent of supply. However, a much higher proportion of supplies of copper, steel and aluminum is provided by scrap.

Import dependence is sometimes confused with vulnerability of supply of minerals. Foreign supplies of most nonfuel minerals are available except in the case of an all-out war in which large areas of the world are engaged in military conflict or transportation routes are

Figure 1.2   *US Net Import Reliance on Selected Minerals and Metals*

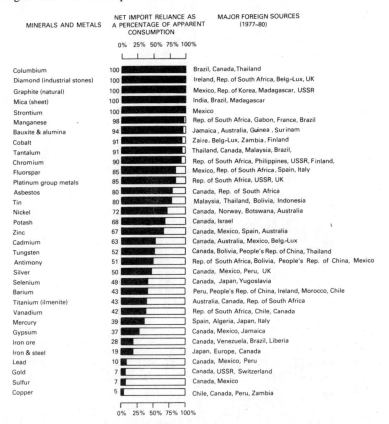

Note: US net import reliance on selected minerals and metals as a percentage of consumption, 1981. Sources shown are points of shipment to the USA and are not necessarily the initial sources of the materials. Net import reliance = imports − exports + adjustments for government and industry stock changes. Apparent consumption = US primary + secondary production + net import reliance. Substantial quantities of rutile, rhenium and zircon are imported, but data are withheld to avoid disclosing proprietary information. Import-export data from the US Bureau of Mines.

Source: Paul R. Portney (ed.), *Natural Resource Policy*, Baltimore, Md: Johns Hopkins University Press, for Resources for the Future, 1982, p. 75.

Figure 1.3 *Net Import Reliance on Selected Minerals and Metals*

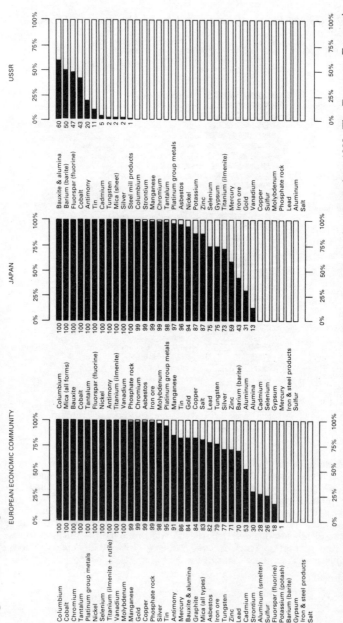

Note: Net import reliance on selected minerals and metals as a percentage of consumption, 1980. The European Economic Community in 1980 included Belgium, Denmark, France, West Germany (FRG), Eire, Italy, Luxemburg, the Netherlands and the United Kingdom. The trade in columbium, tantalum and vanadium is reported together for the EEC. Data for potassium, strontium and graphite are not available for Japan.

Source: Paul R. Portney (ed.), *Natural Resource Policy*, Baltimore, Md: Johns Hopkins University Press, for Resources for the Future, 1982. p. 101.

disrupted. Vulnerability of supplies to disruption during peacetime exists only for a few minerals where production is highly concentrated in a few countries, as is the case with cobalt (Zaire) and with chromium, manganese and platinum-palladium (South Africa).

International trade in nonfuel minerals is exceedingly important to hold down raw material costs in the industrial countries. Fortunately, such trade is relatively free of tariffs and quotas as compared with trade in agricultural and manufactured products. Since a large proportion of nonfuel mineral resources is found in developing countries, the flow of capital and transfer of skills and management to developing countries is exceedingly important for expanding world supplies of mineral products.

Principal Mining Industry Problems and Public Policy Issues

Three categories of problems affecting the world mining industry are addressed in this book. The first concerns problems created by changes in the structure of the mining industry during the past twenty-five years. The second concerns relations between the mining industry and governments, which often reflect conflicts between a variety of social interests on the one hand and the economic interests of the privately owned mining sector on the other. The third category consists of international problems involving both conflict and cooperation at a governmental level.

Structure of the World Mining Industry
The principal change in structure in the world mining industry during the past twenty-five years has been the growth of SMEs from producing a small percentage of the mine output of market economies to producing a substantial percentage in a number of major metals. The growth of SMEs has affected the competitive structure of the world mining industry in three important ways. First, cost elements of SMEs differ from those for privately owned mining firms. Second, the objectives and considerations governing investment decisions of SMEs differ from those of private enterprises. Third, production and marketing strategies of state enterprises tend to be less sensitive to cyclical declines in market demand and price than is the case with privately owned mines.

Most of the large mines owned by SMEs were initially acquired by expropriation of foreign-owned properties for which compensation was paid at a fraction of replacement value. This was true for mining properties in Bolivia, Chile, Guyana, Peru, Venezuela, Zaire and Zambia, among others. In such cases the SMEs began operations virtually free of debt and thus avoided the high interest costs incurred by privately owned mines. Moreover, the large additions to capacities of SMEs in the 1970s and early 1980s were mainly financed by low-interest foreign loans arranged by their governments. These external loans were sometimes provided by international development financing institutions, such as the World Bank and the Inter-American Development Bank, but in most cases the external funds were raised in the private international financial markets with a government guarantee. These capacity expansions financed by low-cost foreign loans are in part responsible for worldwide overcapacity in important metals such as copper and iron ore. In addition to low-cost financing, SMEs have cost advantages over private firms in the form of lower taxes and government provision of infrastructure, such as highways, ports and power, that must usually be supplied directly by privately owned mining firms.

Investment decisions by SMEs are often made on the basis of promoting employment and regional development or of increasing foreign exchange earnings for the country rather than on the basis of relative profit-earning opportunities. SMEs are also not subject to the same financial and political risk considerations as foreign private enterprises. Some SMEs have been developed mainly to supply the domestic market in place of imports which are restricted by means of tariffs or quotas. Any excess of domestic production over domestic requirements is sold abroad, often at prices well below costs.

When the vast bulk of the world's metals was produced by multinational firms, downward adjustments in production took place with cyclical declines in consumption and product prices. However, SMEs tend to be insensitive to price declines in their production and market strategies for two reasons. First, labor costs in developing countries are more a fixed cost because of termination pay regulations and government policies to maintain employment. Second, state enterprises generally seek to maintain exchange earnings in the face of low prices despite the fact that their current receipts may not cover total foreign exchange and domestic currency costs. Privately owned mines in developing countries are also under government pressure to maintain employment. In addition, foreign-owned mines in developing countries have been financed with a high proportion of external debt so that production must be maintained to service debt. Even Western European governments deter mining companies from shutting down or reducing employment in periods of low prices.[5]

The existence of a large segment of the world mining industry in which investment and production/marketing decisions are made more on the basis of government policy objectives than on the basis of private profit maximization has made investment decision-making in the private mining industry exceedingly difficult. Comparative cost advantage and projections of world demand and supply balance no longer serve as reasonably reliable guides for decisions to invest in new capacity. Nonmarket factors, such as subsidized loan financing and a variety of governmental operating subsidies, greatly distort mining costs among countries. SMEs in a number of copper producing countries, including Chile, Peru, Zaire and Zambia, hold large undeveloped orebodies which they plan to exploit as soon as financing can be provided, but they are unlikely to be guided by the projected long-run global demand for and supply of copper. Cyclical fluctuations in mineral prices are exacerbated by the failure of SMEs to adjust output to demand conditions, thereby greatly increasing risks for private investors. Finally, political risks of foreign investment in developing countries have increased, even where such investment is welcome.

Government Mining Policies and Regulations in Developed Countries
Any reader of current mining journals will be impressed by the amount of space devoted to government policies and regulations or proposed regulations that affect the mining industry. Governments have a strong national security interest in the mining industry and the industry has strong support from politicians representing the mining regions of the country. In recent years mining operations in developed countries have come into conflict with a number of environmental interests, including the preservation of public lands for recreational use and ecological balance, and the maintenance of uncontaminated air, water, soil and landscape. Although this conflict between mining and environmentalists first surfaced in developed countries, it has spread rapidly to developing countries as well. Some recent mining contracts between multinational companies and host governments contain rules on environmental contamination that are as strict as comparable mining regulations in developed countries.

The US mining industry is faced with serious problems in meeting environmental regulations, particularly with respect to sulfur dioxide (SO_2) emissions of smelters, and these problems may increase with the imposition of new regulations designed to deal with sulfur sources of acid rain. Compliance with environmental regulations involves large capital expenditures and higher operating costs, which affect the international competitive position of the domestic industry. Other developed countries have comparable SO_2 emission standards, but their administration appears to be less burdensome on the industry than that in the USA.

In most countries outside the USA and Canada, mining industries have been recipients of a variety of government subsidies and domestic markets have been protected by import restrictions. However, the USA has low tariffs and no quota restrictions on primary metals. The US mining industry argues that subsidies on foreign production plus the importance of a strong domestic industry for national defense reasons justify government measures to assist the domestic industry. But the US mining industry has had much less success in lobbying for import controls on minerals than have the more labor-intensive industries such as the textiles. The USA does have antidumping and countervailing duty laws, and there is a real possibility that nonferrous metal imports may be subject to the same types of restrictions, including voluntary export restrictions, that have been applied to steel.

National security concerns for vulnerability to import disruption of strategic minerals have been an important element in US minerals policy, much more so than in other developed countries. Although the USA has resorted mainly to stockpiling imported minerals for dealing with potential import disruption, there exists a strong movement in this country to subsidize the production of relatively low-grade deposits of cobalt, chromium and platinum-palladium for this purpose.

International Mineral Issues

There are a number of international mineral issues in which the world mining industry has a vital interest. Given the competition in mineral products between developed and developing countries, it is not surprising that many of these issues have arisen from differences in organizational structure and strategies between the mining industries in developed and developing countries. One issue has to do with international financial assistance to the mining industries of developing countries. US and Canadian firms argue that such assistance subsidizes the operations of their competitors in the developing countries and the US government has recently opposed loans in support of mining projects in such countries.

Another issue concerns proposals to organize international commodity agreements in copper, iron ore, bauxite and other metals for the purpose of influencing world prices. For more than two decades the governments of Third World countries have sought through the United Nations Conference on Trade and Development (UNCTAD) to negotiate such arrangements, while both private mining interests and governments of developed countries have been generally cool to international agreements for controlling world prices of minerals.

The negotiation of the Law of the Sea Treaty (LOST) brought into conflict the positions of the USA and certain other developed countries with those of Third World countries regarding the control of exploration and development of manganese nodules on the ocean floors. Consortia of US and other developed country mining enterprises have spent hundreds of millions of dollars investigating this source of minerals for eventual development by multinational mining companies, while Third World countries insist that these resources belong to all countries and that exploitation should be governed by an international organization. Also, countries producing substantial amounts of land-based nickel (Canada) and cobalt (Zaire) have a special interest in preventing the mining of manganese nodules from flooding world markets for these minerals.

Finally, the world mining industry has an interest in the formulation of international trade rules governing world trade in minerals. There is considerable danger that trade in nonferrous metals may become subject to the kind of market-sharing arrangements that have characterized trade in steel products. Uncertainties regarding the outcome of these international issues add to the difficulties of investment decisions on mining projects for firms in both developed and developing countries.

Notes

1 Resources in the reserve base include measured reserves plus indicated producible resources for which information is not sufficient for reserve classification. See Bureau of Mines, 1986a, pp. 182–5.
2 Callot (1981) estimated values at the first stage of normal product marketing, i.e. crude oil (not refined products) and concentrates (not refined metals). For the main metals (except iron ore), values represent a certain percentage of the value of the metal content of the ores. Building and construction materials such as sand, gravel and stone were excluded.
3 Because the real prices of most major nonfuel minerals were lower in 1978 than in most years during the 1970s and early 1980s, these percentages were slightly higher in other recent years.
4 The World Bank includes South Africa as a developed, but not an industrial, country and classifies China as a developing country. UN documents usually include both China and South Africa in the developing country category. Some statistical sources put both the Soviet Bloc countries and China together with certain other countries in the category of 'planned economy' countries. In this book South Africa is included as a developed country and China as a communist country.
5 For a discussion of the effects of government policies and large indebtedness on mining operations see Stobart (1984, pp. 259–66).

Appendix Table 1.1a

Major Multinational Firms Producing Nonfuel Minerals, 1980

Firm	Location of Principal Mining Investments:[d]	Principal Nonfuel Minerals	Total assets (billions of dollars)
US Companies:[b]			
AMAX	USA, Australia, Canada, Zambia, Botswana, South Africa, Philippines, Indonesia, Papua New Guinea, UK, Mexico, Dominican Republic	Molybdenum, copper, iron ore, lead, zinc, nickel, tungsten, silver, aluminum	5.1
Aluminum Co. of America (Alcoa)	USA, Guinea, Australia, Jamaica, Dominican Republic, Brazil, Surinam, Mexico	Bauxite, alumina, aluminum	6.0
Anaconda (sub. of Atlantic Richfield)[a]	USA, Mexico, Australia, Chile, Jamaica	Copper, silver, gold, zinc, aluminum, nickel, molybdenum	21.6
ASARCO (American Smelting & Refining Co.)[a]	USA, Mexico, Australia, Peru, Canada	Copper, silver, lead, zinc, gold, molybdenum	2.2
Amoco (sub. of Standard Oil of Indiana)	USA, Canada, Australia, Papua New Guinea	Copper, molybdenum, other metals	24.3
Exxon[a]	USA, Canada, Spain, Chile, Australia	Copper, other base metals	62.3
Freeport-McMoRan	USA, Indonesia, Canada	Sulfur, gold, copper, nickel	1.7
Getty Oil[a]	USA, Australia, Canada, Chile	Copper, other base metals	9.9
Hanna Mining	USA, Canada, Brazil, Guatemala	Iron ore, nickel, bauxite, alumina, aluminum	0.5
Kaiser Aluminum & Chemical	USA, Jamaica, Canada, Australia, New Caledonia, Ghana, Western Europe, Bahrain	Bauxite, alumina, aluminum, nickel	3.6
Kennecott (sub. of Standard Oil of Ohio)[a]	USA, Canada, Australia, Mexico	Copper, molybdenum, gold, silver, lead, zinc, ilmenite	16.0
Newmont Mining	USA, Canada, Peru, South Africa, Chile, Indonesia, Philippines	Iron ore, cobalt, silver, gold, copper, nickel, lithium, molybdenum, vanadium	1.9

Firm	Location:	Minerals	Total assets
Phelps Dodge	USA, Canada, Peru, South Africa	Copper, silver, gold, palladium	2.0
Reynolds Metals	USA, Jamaica, Haiti, Canada, Brazil, Philippines, Ghana	Bauxite, alumina, aluminum, fluorspar	3.3
St Joe Minerals (sub. of Fluor Corp.)[a]	USA, Chile, Australia, Peru, Argentina	Lead, zinc, gold, copper, iron ore, silver	4.7
US Steel	USA, Canada, South Africa, Gabon	Iron ore, manganese, zinc	19.4
Australian Companies:[c]			
Broken Hill Pty (BHP)	Australia, Papua New Guinea, Indonesia	Iron ore, copper, tin, lead, zinc, gold, alumina, aluminum, manganese	6.9
Conzinc Rio Tinto (CRA) (sub. of Rio Tinto Zinc, UK)	Australia, Papua New Guinea, Malaysia	Copper, zinc, tin, gold, lead	4.5
Belgian Companies:[c]			
Union Minière	Belgium, Canada, Brazil, USA, Mexico, Spain	Copper, zinc, silver, gold, platinum-palladium	n.a.
Canadian Companies:[c]			
INCO	Canada, USA, Indonesia, Guatemala, Australia, Brazil, New Caledonia, Mexico, Philippines	Nickel, copper, gold, silver, platinum-palladium, cobalt, magnetite	3.4
Cominco	Canada, Australia, USA, Greenland, Japan, Philippines, Western Europe	Lead, zinc, silver, gold	2.7
Noranda Mines	Canada, USA, Australia	Copper, gold, silver, lead, molybdenum, cobalt	4.6
Falconbridge Nickel	Canada, Norway, Dominican Republic	Nickel, copper, cobalt, iron ore, gold, silver	1.1
Aluminum Co. of Canada (Alcan)	Canada, Brazil, Jamaica, Guinea	Bauxite, alumina, aluminum	6.6

Firm	Location:	Minerals	Total assets
French Companies:[c]			
Pechiney-Ugine Kuhlman (owned by French govt)	France, Greece, Guinea, Australia, Spain, Netherlands, Canada, Cameroon	Bauxite, alumina, iron ore, aluminum	4.5
Le Nickel-SLN (sub. of Elf Aquitaine (owned by French govt)	France, New Caledonia, Cameroon	Nickel	n.a.
IMETAL	France and other Western European countries, Brazil, Morocco, Peru, Australia	Zinc, lead, silver, copper, other metals	1.7
German Companies:[c]			
Metallgesellschaft AG	Germany, Australia, Canada	Lead, zinc, tin, tungsten	1.8
Preussag	Germany, Canada	Zinc, lead, copper, silver, mercury	1.3
Japanese Companies:[c]			
Mitsubishi Metal Corp.	Japan, Australia, Canada, Peru, USA	Copper, nickel, silver, gold, tin	1.5
Netherlands Companies:[c]			
Billiton International Metals (sub. of Royal Dutch Petroleum)	Netherlands, Australia, Canada, Brazil, Surinam, Indonesia, Thailand, Peru, Western Europe, Colombia	Copper, nickel, tin, bauxite, tungsten, zinc, molybdenum	15.5
Patino NV	Netherlands, Canada, Brazil, New Caledonia, Australia, Malaysia, Nigeria	Tin, nickel, cobalt, lithium	0.2
South African Companies:[c]			
Anglo-American Corp. (AAC)	South Africa, USA, Canada, Botswana, Brazil, Zambia, Swaziland, Zimbabwe, Western Europe, Australia	Copper, nickel, iron ore, platinum, manganese, tin, tungsten, diamonds, chromium, silver, zinc, vanadium, gold	4.1

Firm	Location:	Minerals	Total assets
DeBeers Consolidated Mines	South Africa, Botswana, Namibia, Mexico, Lesotho	Diamonds	4.8
Swedish Companies:[c]			
Granges International	Sweden, Liberia, Canada, Saudi Arabia	Iron ore, phosphates	1.2
Swiss Companies:[c]			
Swiss Aluminum Ltd (Alusuisse)	Switzerland, Australia, Canada, USA, Sierra Leone, New Zealand, Guinea	Bauxite, alumina, aluminum, fluorspar, lead, zinc, copper, phosphates	5.0
United Kingdom Companies:[c]			
Rio Tinto Zinc (RTZ)	UK, Australia, Papua New Guinea, South Africa, Canada, Indonesia, Spain, Zimbabwe, Panama, New Zealand	Copper, lead, zinc, iron ore, alumina, aluminum, bauxite, silver	9.0
Consolidated Gold Fields	South Africa, Australia, USA, Philippines, Papua New Guinea, UK	Gold, copper, iron ore, platinum, silver, tin, ilmenite, titanium	2.1
Selection Trust (sub. of British Petroleum)[a]	South Africa, Australia, USA, Sierra Leone	Copper, iron ore, gold, other metals	42.5

[a] Bulk of assets in petroleum or other industries outside nonfuel minerals

[b] Information on US assets taken from *Moody's Industrial Manual*, New York: Moody's Investors Service, 1983 (1982 figures)

[c] Information on foreign company assets taken from *Moody's International Manual*, New York: Moody's Investors Service, 1982 (1981 figures) and "International 500," *Fortune*, August 22, 1983 (1982 figures)

[d] *Engineering and Mining Journal International Directory of Mining*, New York: McGraw-Hill, 1981, and company annual reports for location and minerals

2

Business Strategies in the Mining Industry

Introduction

Business strategy in the mining industry encompasses the same kinds of goals and methods of achieving them as in other industries. The goal may be maximum payout to the equity holders; maximum growth of assets; stability of earnings; eventual liquidation or sale for the highest capital gain to the owners; or some special goal of the managers (as opposed to the owners), such as technological achievements, management security or financial reward. The tactics employed for achieving one or more goals include vertical or horizontal integration; product or geographical diversification; project creation; asset acquisition by merger; expansion through retained earnings or debt financing; the introduction of new technology; and the development of new products. However, mining has certain characteristics that influence both the goals and the tactics for achieving them. First, orebodies are where you find them. Unlike manufacturing, marketing, or service industries, mine location is determined by geology. This explains why mining firms were the first modern nonfinancial multinational corporations. Large mining firms send teams of geologists all over the world to locate new sources of minerals.

A second characteristic of modern mining is that it is highly capital intensive and requires a long gestation period following the initial investment before the product can be produced and sold. This influences the way new ventures are financed and explains why most large mining corporations initiated early in the century were financed by investment houses willing to provide large amounts of venture capital. Later in the post-war period following the Second World War, foreign mining ventures inaugurated by USA and British firms were heavily financed by project loans from international consortia of commercial banks and by government export financing and guaranteeing agencies, such as the US Export-Import Bank. Thus in one way or another, financial institutions have played a major role in the creation of the modern mining industry.

A third characteristic is that mining is a high-risk industry. Risk arises from several sources. First, mineral prices are subject to a high degree of fluctuation. Second, not only is exploration exceedingly high-risk, but there are great uncertainties regarding cost and completion time during the construction period, and in the performance of mines and associated metallurgical plants following their construction. Finally, there are political risks, especially with respect to investments in developing countries.

A fourth characteristic is that most metals are more or less homogeneous products sold in world markets at prices determined on the commodity exchanges, as contrasted with differentiated manufactured products. This means that profitability in mining depends on reducing or limiting costs by introducing better technology and improving management. Marketing and developing new products play a lesser role in traditional metals than in manufacturing. An exception is the development of new metals and alloys and their industrial applications; this is a specialized branch of the materials industry into which some mining firms have entered in recent years.

A fifth characteristic of mining is that every orebody is a depleting resource, the output of which tends to decline over time. Therefore mining firms must continually discover or acquire new orebodies to maintain a relatively stable output over time.

Finally, mining is subject to more regulations and government participation than is manufacturing or most service industries. In many developed countries most exploration and mining occur on public lands, and in many developing countries all minerals in the subsoil are owned by the government. Government policies, therefore, greatly influence business strategies in the mining industry. This subject is dealt with in Chapters 4 and 6.

The above characteristics of the mining industry influence business strategies in a number of ways. Due to uncertainties in exploration, geographical location of mines may be more a matter of taking advantage of opportunities than of conscious planning. Since exploration is usually not undertaken for only one mineral, product diversification or concentration may also be more a consequence of discovery than planning. The desire to take advantage of opportunities may also affect the financial strategy of a mining company. A company that has followed a policy of financing mainly through retained earnings may suddenly find it necessary to incur a large debt in order to develop a newly discovered orebody. This occurred in the case of Freeport Minerals when it developed the Ertsberg mine in Indonesia (see Chapter 5).

The depleting nature of mines has necessitated a dynamic strategy in mining. Not only must companies replace depleting orebodies, they must look for opportunities that may involve products other than those they have been mining. Also, orebodies usually contain byproducts such as gold, silver and molybdenum. These factors tend to induce both product diversification and growth. Therefore a mining firm cannot indefinitely continue traditional operations and survive.

Since a high proportion of world mine output is produced by multinational mining corporations (MNCs) our analysis of business strategy in large part concerns those MNCs. Treatises on the theory of the multinational firm are largely oriented to manufacturing MNCs rather than to those operating in the resource industries (Dunning, 1981; Caves, 1981). The special characteristics of the mining industry outlined above require modifications in the theory of the multinational firm when applied to mining. Except for firms integrated into fabrication, such as aluminum and steel, foreign investment is undertaken to produce for a competitive world market rather than for realizing a market advantage. In the case of the vertically integrated MNC, the motivation for foreign investment is low-cost production of minerals for downstream operations. The location of production is determined by discoveries resulting from worldwide exploration plus consideration of the economic and political climate in the host country.

Product differentiation or product quality plays little or no role in marketing. Firm-specific advantages are mainly found in the professional and managerial resources of the firm. Technology is for the most part common to all large mining operations, including SMEs. There are, however, a wide variety of business strategies involved in financing foreign operations, in the selection of countries in which investments are made, and in whether and how much to produce in areas outside the home country.

Investment in foreign mining operations takes many forms and in some cases does not involve construction and operation of mines. Firms engaged in steel production or in smelting and refining operations may make loans, coupled with long-term supply contracts, to foreign firms producing ores and concentrates, or they may take minority equity positions, coupled with contracts to purchase a certain amount of output, in foreign producing firms. Such arrangements are employed by Japanese steel companies and nonferrous metal processing or fabricating firms, and by international aluminum companies. The operating company is often an SME, as in the case of Companhia Valle do Rio Doce (CVRD) (a major Brazilian state metals producer). These arrangements constitute an alternative to acquiring minerals on the world markets.

Vertical Integration

Mineral firms exhibit varying degrees and forms of vertical integration involving the stages of production from mining ore through the several grades of processing to the production of metals or alloys, and further to the fabrication of basic forms (such as copper wire or aluminum sheets) and finally to the manufacture of finished products. Some firms tend to be more vertically integrated than others. For example, many copper mining companies produce only concentrates which are shipped to other companies for smelting and refining. On the other hand, most aluminum companies are integrated backward into production of bauxite and forward into fabrication and manufacturing. Likewise, most large steel companies are integrated backward to ore production and forward to fabrication of basic steel products.

The degree to which a mining firm is vertically integrated depends in part on the volume of mine production and in part on the availability of financing. Integration of a copper mine into smelting and refining requires a large volume of concentrates and a substantial capital investment. The degree of vertical integration is also determined by business strategy. Having a smelter near a mine saves transportation costs and avoids the possibility of a shortage of smelter capacity, which is usually accompanied by high fees for custom smelting or lower prices for concentrates if sold to smelters. Locating a refinery near a mine is not important for saving transportation costs since the metal content of blister copper produced by a smelter is nearly the same as that for refined metal. It is frequently more important for a refinery to be near the market for the product than near the mine. Thus a number of copper mines have smelters, but even large companies often do not have refining facilities. It is easier to market refined metal than, say, concentrates since there are only a few buyers for concentrates while there is a large world market for refined metal. The advantages of integration through the production of refined metals must be balanced against the high investment cost of smelters and refineries. Even large firms, such as Kennecott Copper Company, delayed forward integration beyond production of concentrates for many years.

Integration into fabrication also presents advantages and disadvantages. Prices of fabricated products fluctuate less than prices of mine products. Integration through fabrication makes it possible to adjust the flow of upstream materials to the market for the fabricated products. On the other hand, fabrication requires a special set of engineering and marketing skills that differ greatly from those possessed by a mining firm. This is one reason why firms that have integrated forward into fabricating frequently acquire going concerns rather than building fabricating facilities.

Firms specializing in a particular stage of production may also integrate backward into mining. This occurred in the case of several large US steel companies that integrated into the mining of iron ore. ASARCO, which originally specialized in nonferrous metal smelting and refining, integrated into mining when it lost some of its large mining customers as a consequence of their forward integration.

Changes in the structure of markets, government regulations and shifts in product demand have affected the location of vertical integration. Initially the large US aluminum companies were forced to integrate into bauxite production in developing countries because US reserves of bauxite were not sufficient to supply their requirements. As the cost of power rose in the USA relative to the cost of hydroelectric power in countries such as Brazil and Venezuela, aluminum companies began producing more aluminum abroad as well. Increasingly, however, aluminum has been produced by foreign state enterprises so that it has become cheaper to buy foreign aluminum than to supply it from US smelters. Alcoa recently announced plans to reduce domestic aluminum production by 1995 to about 50 per cent of its domestic requirements as compared to 85 per cent in 1985. Alcoa expects to become a purchaser rather than a seller of primary aluminum and to concentrate to an increasing degree on the development of new aluminum-based materials (*Wall Street Journal*, September 1985, p. 7).

During the 1980s several US copper smelters closed because they did not meet Environmental Protection Agency (EPA) pollution abatement standards. Anaconda closed its American copper smelters rather than undertake costly modifications required by EPA regulations and contracted to sell its copper concentrates to Japan. Subsequently it sold its copper mining assets as well. Other US smelting capacity will be forced to close in the future and there is some question whether it will be replaced. If metal prices remain low in the face of high US mining costs, it will be more economical for some US metal companies to acquire their minerals from abroad and concentrate on fabricating and manufacturing.

Horizontal Integration

Horizontal integration has been the traditional strategy for the growth of mining firms. It utilizes the professional skills and managerial experience of the firm and complements its need to acquire and develop new orebodies as existing ones are depleted. The acquisition of mines in the same area may provide economies in the utilization of processing facilities, such as concentrators, or may render economic the construction of a smelter for treating concentrates produced by several mines.

Horizontal integration in mining may take place in several ways. The mining firm may undertake exploration to find additional reserves. Alternatively, companies may acquire orebodies, or the right to develop them, that have been more or less fully explored or even partially developed by others. However, the geologists of the firm acquiring an orebody are likely to undertake considerable exploratory work on their own in order to verify the data of others, or to obtain additional information on the orebody.

Exploration may take place at different stages. Exploration over large areas covering several thousand square miles is called "grass roots" exploration. This takes the form of aerial photogeology and general geological reconnaissance to identify potential orebodies, this is followed by exploratory drilling and other physical examination designed to define the grade, shape and size of the orebody. If such investigation yields positive results, more intensive exploration involving sinking of shafts and metallurgical testing of ore samples takes place.

Exploration leading to the development of St Joe Minerals' El Indio gold/copper mine in northeastern Chile provides an example of several stages of exploration. The existence of the orebody first came to light when a St Joe geologist learned of a mining claim from which high-grade gold ore was being extracted on a small scale and transported by mule from a mine at the 13,000-foot level in the Andes to a Pacific port over a hundred miles from the mine. This information was gleaned from public records of mining claims and conversations with local residents. Following negotiations with the family owning the claim, St Joe undertook both intensive exploration of the claim and extensive exploration of a much wider area in the region of the mine. A Chilean company was formed, Compania Minera el Indio, which is owned 80 per cent by St Joe Minerals and 20 per cent by the family holding the original claim. On the basis of a feasibility study started in 1978, a decision was made in 1979 to develop a modern mine to produce both gold and copper. The mine has proved to be one of the most profitable in the world.

Another horizontal integration strategy is to acquire ownership and control of an existing mining firm. If the acquired firm also has undeveloped reserves, the new owner may plan to develop the reserves as well as operate the existing mine. Acquiring a mining company generally means acquiring the services of a team of experienced personnel, including mining engineers, geologists and managers. The skilled workers and professional personnel of a firm are often a major factor in an acquisition decision. It is frequently more economical to acquire ore reserves and mines by taking over going concerns than to obtain reserves by exploring and building new facilities to develop them. If the market value of the equity shares of a company is low because of low product prices or managerial inefficiency, the asset value of the firm may be far higher than the market value of the shares. It may also be easier to finance the acquisition of a going concern than to acquire an equivalent amount of assets by exploring for reserves and constructing a new mine. For example, controlling interest in a mining firm might be obtained with the equity shares of the acquiring firm.

Large mining companies tend to employ a combination of strategies for acquiring additional reserves. Most of them have assembled teams of geologists that conduct exploration both within the home country of the firm and throughout the world. These geologists not only undertake direct exploration, but investigate the findings of other geologists with a view to determining whether more intensive geological work is warranted. A team of first-rate geologists is a valuable asset and a company may want to keep them busy discovering new opportunities, even if it has no immediate plans to develop new orebodies. Ore reserves can often be held for considerable periods of time without being developed and development can be timed in relation to the outlook for the price of the products and the financial and human resources available to the firm. Geologists may also investigate the reserves of other firms and thereby discover opportunities for taking over these firms. Thus, the strategy employed by a firm in horizontal integration may be determined in part by the nature of the opportunities that come to light.

When a mining firm sends geologists abroad to do grass roots exploration or investigate discovered orebodies, the firm may be led into geographical diversification. Before a company invests substantial funds for exploration in a foreign area, it will take into account the political and economic conditions for doing business in that country. However, there is usually no prior decision on the part of management to make an investment in one country over another. In some cases the government or the geological survey office in a particular country will invite geologists from a number of multinational mining companies to review the geological work that has been done. This may lead to contracts with several multinational firms to undertake exploration of certain areas of the country over a specified period of time. Such exploration contracts usually include the right to obtain an exploitation contract over a portion of the original exploration area if an orebody is discovered. The actual development of a mine would depend upon more intensive exploration and the outcome of a feasibility study.

Product Diversification

Product diversification by a mining company occurs in several ways. First, many orebodies contain more than one economically producible mineral. For example, a copper mine may contain a certain amount of gold or silver or molybdenum as a byproduct. Whether it will pay to recover the byproducts frequently depends on the size of the mining operation, since special processing is required for each byproduct. In some cases the development of a copper or nickel mine would not be profitable without processing the byproducts. A second way in which product diversification occurs is through the discovery of an orebody that contains a mineral that is not a principal product of the firm. For example, a mining firm may decide to develop a silver orebody its geologists have discovered even though the principal product of the firm is nickel. Product diversification may also result from a deliberate decision on the part of a mining firm to diversify into other products, in which case it may instruct company geologists to explore in areas known to contain the mineral(s) into which it decides to diversify. Alternatively, the firm may acquire an orebody containing the target mineral that has been partially or fully explored by another firm.

An example of the discovery of a mineral that is not the principal product of the company is that made by Standard Oil of California (Socal) geologists. While exploring for petroleum in Colombia, they discovered an orebody containing iron and nickel. Socal invited Hanna Mining to enter into a partnership to explore the orebody, which was eventually developed as a nickel mine by the joint venture, Cerro Matoso. In this case, Socal decided to maintain only a small equity interest in Cerro Matoso. However, had Socal decided to diversify into nickel, it might have become a major equity holder in the venture. As matters developed, the subsidiary (Billiton) of another oil company (Royal Dutch Shell) later became a major equity holder in Cerro Matoso (Mikesell, 1983, Chapter 8).

One of the principal reasons for diversification is risk reduction. The long-run demand for a particular metal may decline, so the risk is reduced by producing several metals. Product diversification may also improve the stability of a firm's earnings when prices of two different metals exhibit countercyclical patterns. For example, historically when base metal prices are low, gold prices rise; a producer of both copper and gold could expect to have more stable revenue than a producer of either metal exclusively. This is illustrated in Figure 2.1. On the other hand, if there is a high positive correlation of price movements between two products, say, gold and silver, diversification into production of both metals will lead to greater

Figure 2.1   *Copper and Gold Portfolio Effect*

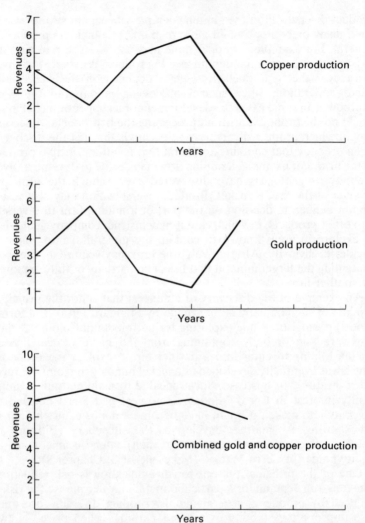

instability of revenue than if production were concentrated exclusively in either gold or silver. This is illustrated in Figure 2.2.

Nonfuel mineral companies may also diversify into fields such as oil or coal or even into manufacturing or services that have little relation to mining and mineral processing. Kennecott Copper Company acquired Peabody Coal, and AMAX and Newmont Mining both diversified into petroleum. Again the reasons were risk reduction and

Figure 2.2   *Silver and Gold Portfolio Effect*

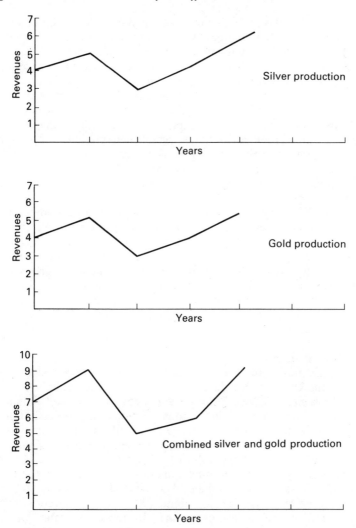

stability of earnings. Thus, large mining companies frequently become conglomerates. Diversification into production that requires professional skills and managerial expertise unrelated to a firm's traditional business has often proved unprofitable. One reason may be that a firm is unable to utilize in another industry the managerial talents and experience that led to its success in a particular industry.

Financial Strategies

Financial strategies in mining have changed with the evolution of the industry. During the nineteenth and early twentieth centuries nearly all successful mines originated with a prospector's discovery based on mineral outcroppings. The prospector staked a claim, but except for small gold and silver mining operations, he was usually unable to finance intensive exploration and development. Commercial bank loans were unavailable in the absence of collateral. The owner of the claim had to obtain venture capital from wealthy individuals or negotiate an agreement with an established mining company for exploration and development of the discovery. In the latter case, the owner of the claim received a percentage of the net income or shares in the mine if it were incorporated, but lost control of the property. Many large mining companies were developed by acquiring claims in this way. With the disappearance of easy-to-find orebodies, most discoveries are now made by teams of professional geologists associated with established mining companies.

In considering methods of mine financing, a distinction must be made among three categories of mines: small mines with assets of under one million dollars; medium mines with assets of less than one hundred million dollars; and large mines with assets running from several hundred million to over a billion dollars (Mikesell, 1986b, pp. 2844–9).

Small-Scale Mines
Small-scale mines are variously defined and what is regarded as small-scale in the USA or Canada in terms of capital cost would be large in countries such as Chile or Bolivia where there are a substantial number of mines producing less than 500 tons per day. Using the definition of a small-scale mine as one producing under 100,000 tons annually, it has been estimated that these mines produced 16 per cent of the total world value of nonfuel minerals in 1982 (Carmen, 1985, pp. 119–24). In the case of several nonfuel minerals, over half the output is produced by small-scale mines. They include beryllium (100 per cent); chromite (50 per cent); graphite (90 per cent); mercury (90 per cent); and tungsten (80 per cent). In terms of value output, substantial amounts of copper, gold, tin and iron ore are produced in small-scale mines, but these amounts generally represent less than 10 per cent of the total value output of the individual minerals mentioned. In the USA, Australia and Canada, as well as in the Latin American countries, there are large numbers of small mines, the ownership of which is held by a family or a few individuals who supplied the initial equity capital. In the USA and Canada, small

mines are mainly limited to the production of gold and silver. Metals such as copper, nickel and iron ore require large amounts of capital and are produced almost entirely by mining companies with large assets. Exploration and development of small mines are typically financed by equity; debt financing from commercial sources is almost impossible to obtain. About the only type of debt financing available is for equipment purchases. In such cases the equipment serves as security for the loan in addition to a repayment guarantee by the borrower. Banks generally refuse to lend to small mining ventures and, in any case, the risks and the long period before there is any positive cash flow make debt financing imprudent.

Most small mines are financed by individuals or partnerships. In a typical scenario, the owner of an inadequately explored claim is joined by a promotor who attempts to find individuals with financial resources who are interested in mining. The promoter convinces a group of investors that a fully operational mine can be built for a relatively modest amount ranging from $50,000 to $250,000, depending upon the promoter's perception of how much financing can be raised by the group. This is done without benefit of a geologist or mining consultant. After venture capital has been assembled and a lease or development agreement negotiated with the claim owner, the promoter goes ahead with construction. As construction proceeds, the investors are induced to commit additional funds to save what they have already invested, since the initial funds are almost never large enough to complete the mine. This process may continue until the project has consumed two or three times the initial amount before the investment group realizes that it may never own an operating mine. At this point the group may hire a mining consultant. In most cases he will advise that the project is not feasible and no more funds should be contributed. This sequence is fairly typical of projects initiated in the USA and Canada. In a minority of cases, the project will be picked up by a mining company with resources available to develop a profitable mining operation.

It has sometimes been possible to raise funds for a mining venture with a public stock offering or with a private placement of stock. This method of financing was employed during the rapid growth of "penny" stock markets in 1979–82 when a number of public stock issues were floated in western cities such as Denver, Spokane and Vancouver, BC to finance small mining. Some of these projects were technically sound while others were not. For example, a modestly successful company, The Silver State Mining Company of Nevada, was funded through a public offering in 1981. This company was formed by two brothers; they were geologists who had acquired a small gold deposit. The mine was constructed on a "shoestring" budget and after three years of operation it still had not reached full productive capacity and the company had not made any profit. The company then sold a 60 per cent interest in the mine to another company and the joint venture partner provided financing to complete the mine. The company will probably be successful, but there has been little appreciation of the shares held by the public stock investors.

During the early 1980s penny stock offerings for financing mining ventures in Canada, the western United States and South America were made on the Vancouver, BC stock exchange. The stock offerings were in amounts up to $500,000. The most successful projects were promoted by geologists and mining engineers.

Medium-Size Mining Companies

Medium-size companies typically have income from several mining operations. Many of the companies in this group were once small and have become joint ventures, or the owners have farmed out their projects to larger companies for management and development. Companies in the medium-size group are usually able to obtain commercial bank financing for their projects.

Although the number of medium-size mining companies in the USA is small, they are numerous in Canada. A portion of the equity for the development stage of the projects may have been provided by one or more large mining companies. These companies obtain options to buy additional shares in the event the project is successful. The projects typically do not utilize debt financing although occasionally a portion of the capital will be provided by a project loan. It is more common to use debt financing to expand the capacity of a mine once it is in operation.

Small and medium mines are the norm in many other countries, including Mexico, Peru, Chile, Australia, Finland and the Philippines, but are less common in South and Central Africa. Several South American governments have organized special investment banks for financing small and medium mines.

Large Mining Companies

The major US copper companies founded in the late nineteenth and early twentieth centuries required large amounts of capital to develop open-pit porphyry orebodies, most of which are located in Utah, Arizona and New Mexico. Local financing for these projects was not available. Much of the financing came from wealthy Eastern investors who had an interest in mining; from the owners of successful gold and silver mines; and from exporters of copper who sold American copper to Europeans. Some of these same sources financed the development of large-scale mines in Chile, Mexico and Peru, which became affiliated with large American companies, such as Anaconda, Kennecott and ASARCO.

Mining companies have traditionally avoided a high debt-equity ratio for two reasons. First, mineral prices tend to fluctuate widely, and second, investment in new mines requires a long gestation period. Debt service payments may become due immediately, while income from the new mining or metallurgical facilities may be long delayed and subject to substantial fluctuation.

In the case of a mining firm that is owned by an individual or a family or closely held by a few large stockholders not in need of immediate cash flow, expansion by means of reinvested earnings may be a desirable strategy. This strategy avoids the cost of borrowing or of issuing new equity shares that dilute control by existing investors. Since taxes on capital gains are substantially less than taxes on dividends, the owners may find it to their advantage to limit dividends and increase the value of their shares in the mining firm. On the other hand, where shares of a mining firm are widely held a high payout increases the market value of the shares; this facilitates expansion by means of issuing new equity shares or by using shares in the company to take over other companies.

Prior to the 1960s, most large American mining companies maintained a high payout of dividends to equity holders and long-term debt tended to be relatively small or nonexistent. Since these companies were quite profitable, it was easy for them to raise funds by issuing additional stocks to existing equity holders or exchanging their stock for that of smaller companies. For example, over the period 1957–63 both Kennecott and Phelps Dodge paid dividends equal to 80 per cent of their after-tax profits. In 1967 neither Phelps Dodge nor Newmont had any long-term debt and Kennecott's debt-equity ratio was quite low. Of the six major American mining companies – Phelps Dodge, Newmont, Kennecott, ASARCO, Anaconda and AMAX – only AMAX had a substantial debt-equity ratio (38 per cent) in 1967. However, the long-term debt of these companies increased substantially after 1965, in part because of their diversification into industries other than nonfuel mining, e.g. aluminum smelting, petroleum, uranium and coal, and in part because of their need to finance investments in environmental controls required by the EPA.

As is shown in Figure 2.3 prior to 1965 US and Canadian nonferrous metal companies maintained quite low debt-equity ratios in the range of 0.2 to 0.4. After reaching a peak of nearly 0.5, the average debt-equity ratio of these companies declined throughout most of the 1970s. The fall in metal prices reduced the incentive to

Figure 2.3   *Debt-Equity Ratios for Primary Nonferrous Metals Industry in the United States and Canada (1960–81)*

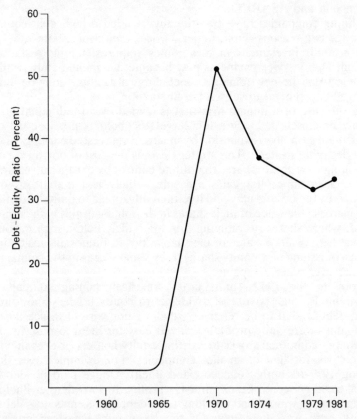

Source: Federal Trade Commission – Securities and Exchange Commission, *Annual Financial Reports Mining Corporation, Value Line Investment Survey*, L. Arthur English of the Toronto-Dominion Bank. (The 1981 update is based on a sample of 25 US and Canadian mining firms.)

invest in increased capacity and the decline in cash flow from operations reduced the ability to cover debt service. Several leading mining companies liquidated assets to reduce debt. By 1985 Anaconda had liquidated nearly all its copper producing assets in the US, while Kennecott liquidated its major noncopper holdings.

Development Patterns and Business Strategies

The development patterns of mineral industries exhibit business strategies peculiar to the markets for the products of the industries. The development of the traditional metal industries – copper, gold, iron ore, lead, silver and zinc – was mine-oriented in the sense that exploration and development occurred in response to the growth of world demand for the metals. Prospectors discovered ore deposits and either raised capital for development or sold their claims to mining companies or investment firms specializing in mine finance. As mining companies grew in size they carried on exploration to increase their reserves. Iron ore development, on the other hand, has been mainly generated by consumers. The large steel corporations established in the second half of the nineteenth century developed their own supplies of iron ore and coal. They did so by taking over existing mining companies or developing both domestic and foreign mines. Thus in both the USA and Europe, the large steel firms are integrated from mining through fabrication. Although nickel is mainly used as an alloy in steel production, nickel mining was developed by mining firms and remains independent of the steel industry. However, since nickel smelting is both technologically complex and very capital intensive, smelting became the dominant function; today mining and smelting are integrated and the industry is dominated by a few large companies carrying on both functions.

Aluminum production is relatively new, dating from the early part of the twentieth century. The first companies produced aluminum metal for sale to fabricators, but in order to expand the market for their products these companies began to manufacture products in the early 1920s, in some cases by buying out their customers. Initially, alumina (the raw material for producing aluminum metal) was produced from bauxite deposits in the southeastern United States, but after the industry was well established, aluminum companies began to develop foreign bauxite reserves.

In some of the newer metals, such as titanium, magnesium and beryllium, technology was the driving force in the establishment of the industry. In the case of titanium, which is required by defense industries, the industry was developed with US government assistance. Beryllium is dominated by a single company which developed the complex technology necessary for production.

Both technology and the increasing capital intensity of production have been major factors in the evolving of the mining industry. The small and medium mines producing copper, lead and zinc in the nineteenth century sold their ore or concentrates to smelting and refining companies. For example, there were some 3,000 copper mines in the USA at the beginning of the present century, few of which had concentrating facilities. However, transportation costs favored integrated operations at least through smelting, and only the larger copper firms had the technology and capital for processing. Also, early in this century both mine technology and metallurgy favored large open-pit mines costing hundreds of millions of dollars. Therefore, little US and Canadian copper is produced in small mines today. A similar development occurred in lead and zinc mining, although there are still a number of medium-size mines delivering concentrates to larger firms for smelting and refining. Small and medium mines continue to be important in gold and silver production, but since these commodities are frequently byproducts of other nonferrous metal production, a significant amount of gold and silver output is accounted for by large-scale production of other metals.

Illustrations of Business Strategies in Mining

As noted in Chapter 1, world (excluding the Soviet Bloc) output of the major metals is dominated by a small number of large MNCs and by SMEs in developing countries. The manner in which MNCs have evolved reveals certain business strategies or combinations of strategies. In several cases, the initial strategy was to acquire several mines or orebodies oriented to a single product. In other cases, companies initially engaged in smelting and refining mine products produced by other firms later integrated backward into mining. In still other cases, the initial orientation was mine financing and equity participation through a holding company. Some mining firms were diversified into several minerals or stages of production, such as refining and semifabricating, early in their history, while others tended to concentrate on a single metal at the mining stage. As metal mining companies grew larger, most of them diversified into other natural resource production, such as coal and petroleum, and into manufacturing. While large mining firms have not followed clearly defined growth paths, the following examples illustrate major strategies employed in the creation of a firm.

Kennecott Copper Company

Kennecott Copper Company, which became a subsidiary of Standard Oil Company of Ohio (majority owned by British Petroleum Company) in 1981, is an example of a company initially oriented to mining a single metal. Kennecott Copper was formed by the union of several copper mines in the western USA, Alaska and Chile that had been financed by American investment firms. The four western US mines, located in Utah, Nevada, Arizona and New Mexico, were all under the managerial direction of Daniel J. Jackling who revolutionized copper mining by introducing mass production methods for exploiting low-grade porphyry deposits. A principal source of funds for development of these four mines was the Guggenheim Brothers, who also advanced funds to develop the El Teniente mine in Chile and (with J. P. Morgan & Company) provided funds for development of a high-grade copper deposit in Alaska near Kennicott Glacier. In 1915 the several financial interests decided to combine their mining companies into one, taking the name (with a change of spelling) of the Alaska company. Subsequently through acquisitions and mergers Kennecott grew to include three other copper mines. Kennecott's exploration subsidiary, Bear Creek Mining Company, discovered additional porphyry deposits in the USA and other countries, but the company chose to expand existing mines rather than to develop its discoveries. They were later developed by other companies.

Kennecott entered the fabricating business by acquiring Chase Brass & Copper Company in 1929 and later acquired an electrical communication wire and cable manufacturing company. Since custom smelters and refineries, such as ASARCO, were in operation, Kennecott did not find it economical to provide its own smelters and refineries until the 1960s. In addition to becoming a fully integrated copper company, Kennecott later diversified into coal, lead, zinc, silver, iron ore and other minerals. For example, Kennecott acquired Peabody Coal, one of the largest US coal producers, but the acquisition was challenged by the US government and the coal company was eventually sold. Many observers believe the large sums spent on diversification would have been better spent on modernizing Kennecott's relatively high-grade copper mines.

Phelps Dodge

Phelps Dodge also specialized in a single metal – copper – but its growth strategy was different from that of Kennecott. Early in the nineteenth century the family firm was in the copper import business and its interest in mining began with the purchase of some Arizona properties in the 1880s. (Phelps Dodge was run as a family partnership until 1917 when it was incorporated.) By the turn of the century Phelps Dodge was producing some 30,000 st of copper annually and completed its first smelter in 1902. Unlike Kennecott, Phelps Dodge integrated into fabricated copper and brass products at a relatively early stage and by the early 1930s was refining most of its smelter output. Thus, Phelps Dodge became America's first fully integrated copper company. Phelps Dodge did little grass roots exploration (before the 1970s), but adopted the strategy of acquiring deposits explored by others. In its downstream operations, Phelps Dodge emphasized copper wire fabrication and became the leading US wire producer.

Beginning in the 1950s Phelps Dodge invested in oil exploration and in aluminum, neither of which proved to be profitable. It also invested in mining companies in other countries, including Peru, Australia, South Africa and Turkey, but does not operate the mines. The bulk of its assets are in domestic copper mining, processing and manufacturing.

Prior to 1969, Phelps Dodge financed its growth from earnings or by issuing stock. From 1969 to 1974 the company accumulated long-term debt totalling $400 million so that its debt-equity ratio rose to 1.0. Also during the 1970s the company switched from a policy of acquisition of known deposits to exploration in the USA and abroad.

ASARCO

In contrast to Kennecott and Phelps Dodge, ASARCO began as a processor of metals mined by other companies. In 1899 it was organized as the American Smelting & Refining Company through the consolidation of a number of US smelters and refineries under the direction of a group of New York financiers. Initially the company's major activities were in lead smelting, but later it moved into copper, zinc and silver processing, through the purchase of other firms and the construction of smelting and refining facilities in several locations in the USA. Although ASARCO entered the copper mining business in 1922 with the organization of the Northern Peru Mining & Smelting Company, followed by purchases of mining interests in Canada, Mexico and Australia, it did not mine domestic copper until 1954 when it bought the Silver Bell Company in Arizona. Kennecott acquired smelter capacity for the output of its Bingham Canyon mine in Utah (the largest copper mine in the USA) by purchasing ASARCO's Garfield, Utah smelter in 1959. ASARCO became concerned whether its smelters at Hayden, Arizona and El Paso, Texas would have enough custom work to operate. (At that time Kennecott's mine output represented 50 per cent of ASARCO's copper smelting business.) Instead of selling these properties, ASARCO moved to assure a source of concentrates by expanding its copper mining operations. By the late 1960s, 75 per cent of ASARCO's before-tax earnings came from mining, while prior to the 1950s the bulk of its earnings was derived from custom smelting and refining operations.

ASARCO integrated into fabrication to a lesser degree than Kennecott or Phelps Dodge. On the other hand, ASARCO diversified into foreign operations to a greater degree than either Kennecott or

Phelps Dodge. It acquired a major interest in the Mt Isa copper, lead and zinc mine in Australia; developed a large copper mine in Mexico; and became the majority stockholder and manager of Southern Peru Copper Company (SPCC) in the early 1950s. SPCC's two mines, Toquepala and Cuajone, are among the largest copper mines in the world. Currently ASARCO's foreign mine capacity exceeds its domestic capacity and it is one of the world's largest copper producers.

Newmont
Newmont, which is America's fourth-largest copper mining company, followed an entirely different pattern of development. Newmont was founded in 1921 as a holding company acquiring minority equity positions in companies in the USA, Canada, Australia and South Africa. These companies produced copper, gold, zinc, nickel, chromite, silver and other minerals. As Newmont increased its equity holdings in several of these companies to majority or 100 per cent ownership, it became directly involved in mining operations. Newmont's wholly-owned and operated companies include (in order of amount of sales) Magma Copper Company (US), Carlin Gold Mining Company (US), Newmont Mines Ltd (Canada), and Newmont Proprietary Ltd (Australia). In addition, Newmont has majority ownership in Foote Minerals Company (US) and the O'okiep copper mine (South Africa). Although most of its properties were acquired through purchase, some (including Carlin Gold Mining) were acquired as a result of Newmont's exploration.

Newmont maintained a relatively small exploration staff and most of its exploration efforts have been focused on gold. Its primary strategy has been to buy into successful companies or to participate in joint ventures. Its mine development activities have been confined to gold mining in Nevada and lithium mining in Chile. Through its affiliate Foote Minerals, Newmont has become the world's largest lithium producer. (Foote also manufactures manganese alloys for the steel industry.)

As of 1980 Newmont's assets totaled nearly $1.5 billion and with liabilities of about $500 million, of which $144 million represented long-term indebtedness. During much of the period of its growth until the late 1960s, Newmont had no long-term indebtedness.

Rio Tinto Zinc
Rio Tino Zinc (RTZ) was built from a minor investment trust to one of the largest mineral companies in the world, with a broad portfolio of investments in copper, iron ore, bauxite and aluminum, uranium, lead, zinc, industrial chemicals, and various manufacturing operations.

The Rio Tinto company was formed with Rothschild financial backing when the Rio Tinto mine in Spain was acquired in 1873, but as a consequence of declining ore grades, by the late 1940s the only significant income of the Rio Tinto company came from its portfolio shares in companies operating in the Northern Rhodesian copper belt. The company was reorganized as Rio Tinto Zinc in 1962, and as a result of a series of investments built a chain of largely autonomous companies involved in mineral production throughout the world. Although there is decentralization and local managerial control of affiliates, significant capital expenditures are subject to RTZ's approval. RTZ's affiliates and associated companies include Conzinc Rio Tinto of Australia (CRA) and its subsidiary, Bougainville Copper Ltd in Papua New Guinea (PNG); the Palaboro Mining Company in South Africa; Lornex Mining Company in British Columbia; Riotinto Minera in Spain; the German metallurgical company, Duisburger Kupferhutte, and minority interests in exploration companies and other mineral ventures involving production of aluminum, uranium, iron ore, lead, zinc, tin, coal and petroleum. RTZ's assets as of 1980 were valued at about $8 billion.

Conzinc Rio Tinto of Australia

RTZ's largest subsidiary, and one of the world's largest multinational mineral firms, has had a unique business history and growth strategy. The original company, the Zinc Corporation (ZC), formed in 1905, initially processed zinc-bearing tailings from an Australian mine. ZC later acquired equity in an Australian smelter and mining properties. As a consequence of mergers arranged by British owners in London, ZC became Conzinc Rio Tinto (CRA) in 1962, the same year RTZ was incorporated. In 1971 CRA and New Broken Hill Consolidated Ltd (NBHC) merged their lead/zinc/silver and other interests into an integrated company, Australian Mining & Smelting Company (AM&S), which in turn operated smelters in Australia, Europe and other areas of the world. CRA also acquired interests in bauxite deposits and in alumina and aluminum production plants, and created the Bougainville copper mine in PNG. Through subsidiaries and joint ventures, CRA also entered into the production of tin, petroleum, uranium, iron ore, coal and diamonds. Although RTZ still holds a majority interest, a large portion of the stock of CRA is held by Australian citizens and management is in the hands of Australians.

CRA has followed an expansion strategy of diversification into mining a number of minerals in different geographical areas followed by integration into processing and manufacturing. The sequence for each mineral has been quite similar. First a deposit was found and following a feasibility study, mine development was financed through a combination of internal and external equity financing. Many of the new corporations formed to develop deposits were joint ventures, minority shares of which were sold to the public to raise additional funds. After a mine was successfully developed, other corporations were formed for processing or fabricating the mined product. CRA also followed a strategy of subsequently buying out its joint venture partners; by doing so it became the controlling shareholder of each affiliated company.

Anglo-American of South Africa
Another multibillion-dollar mining company, Anglo-American Corporation of South Africa (AAC), also developed as a mineral-oriented holding company, beginning with investments in gold mining operations in South Africa and extending to diamonds (through minority ownership of DeBeers), nickel, copper, platinum, uranium, coal, and other minerals in central Africa, Canada and the USA. In most of the companies in which AAC has an interest, there is little direct management control, but it exercises considerable influence over policies and financing.

A major strategy of AAC has been to initiate and finance new mineral projects by acquiring a minority interest in existing firms. Following the formation of an affiliate, it sells some of the shares to other companies and the balance to the public. To maintain a measure of control, it provides management and administrative services to its minority-owned affiliates. For example, AAC acquired control of Hudson Bay Mining & Smelting in Canada. Hudson Bay was then used to acquire Inspiration Copper Company in the USA, which in turn was used to buy large undeveloped zinc reserves in the USA, together with coal mines, small oil and gas companies, and agricultural chemical companies. Once these acquisitions were made, the entity was reorganized with Hudson Bay being merged into Inspiration and the entity renamed Inspiration Resources. Inspiration Resources then made a public offering in the USA and Canada to raise additional capital for operations and further acquisitions. AAC's influence in minority-owned affiliates has been achieved in considerable measure through its professional staff which provides consulting services to affiliates.

Strategies of State-Owned Mining Enterprises

Over the past fifteen years there has been a rapid expansion of SMEs in Third World countries. Together these enterprises now control a substantial proportion of Third World output of bauxite, copper, iron ore, phosphates and tin (see Appendix Table 2.1a). In this section, we are concerned with the business strategies of SMEs and how they differ from the strategies of multinational companies.

Many of the initial advantages of the MNCs were acquired by SMEs when the foreign-owned mines were expropriated and the SMEs inherited a trained, experienced professional and managerial staff. Although they were cut off from technical and managerial services of the parent MNCs, some of these advantages can be provided by international consulting firms, such as Bechtel Civil & Minerals, Inc. and Dames & Moore. Both SMEs and MNCs use the services of international engineering firms to construct new mines and metallurgical plants, conduct training programs and carry out exploration. SMEs use much the same types of equipment as MNCs and can obtain assistance in installing and operating equipment from equipment manufacturers. Most mine products are sold in competitive world markets in which SMEs have been able to establish efficient marketing programs. Finally, until recently SMEs have had an advantage over MNCs in obtaining international capital which they could borrow at favorable interest rates with a government guarantee and sometimes at subsidy rates from international development agencies such as the World Bank.

Generalizations regarding the performance and strategies of SMEs in the mineral industries are likely to be misleading since there are substantial differences among them with respect to objectives and managerial competence. Some have performance records as profit maximizers comparable to the best MNCs, while others have proved inefficient and have pursued goals unrelated to profitability. However, contrary to the opinions expressed by many officials of MNCs, most SMEs have reasonably competent managers that seek to maximize the earnings and growth of the enterprise. In the case of those SMEs that were initiated following the nationalization of a foreign enterprise, the managers, professional and skilled personnel constitute a legacy of the former owners. After a foreign-owned mining subsidiary has operated in one of the more advanced developing countries for a couple of decades, most managerial and professional personnel are natives of the country and tend to be retained by the government owner. For example, at the time of the complete nationalization of Kennecott's El Teniente mine in Chile by the Allende government, there were only five American company employees in the mine. To cite another example, when the Peruvian government expropriated two large American-owned mining companies in the early 1970s (Cerro de Pasco and Marcona Iron Mines), the managing director of Cerro and the president of Marcona were Peruvians. Although the top officials were discharged at the time of expropriation, most Peruvian managerial personnel remained with the nationalized companies.

SMEs are usually managed as separate entities, independent of the bureaucracy of the government. The government appoints the chief executive officer and the board of directors, and provides general policy direction, but operations and much of the business strategy are under the control of managers who have a vested interest in the profitability and growth of the enterprise. Thus large state mining companies, such as Brazil's CVRD and Chile's CODELCO, have much in common with an American or British mining corporation of comparable size. (In 1984 CODELCO and CVRD ranked 28th and 104th among the world's largest mining companies in terms of annual gross revenue.) This is not to say there are not political appointees in some SMEs who regard their position as stepping-stones to political advancement within the central government. But even these officials are likely to be judged in considerable measure by their contributions to the success of the enterprise. Unlike ministries of public works or welfare, the management of a large mining firm requires technical competence and business ability in order to make a profit, or at least to avoid losses that would be a drain on the government budget. Moreover, most mineral SMEs export to a competitive world market so their profitability cannot be assured by a protected domestic market.

Despite the existence of considerable independence, managers of SMEs must be responsive to government policies that may conflict with the business-oriented goals of profit maximizing, risk minimizing and growth. Government policies frequently favor holding down prices to domestic purchasers, buying local materials and equipment, paying high wages and negotiating labor contracts favorable to unions, undertaking uneconomic investments, or using inefficient production methods to promote employment. There are, however, government policy interests that are compatible with the goals of SME managers. All governments want their SMEs to maximize foreign exchange earnings. Although there are cases where government policy favors international barter deals or sales to politically favored countries that may be in conflict with this goal, SMEs usually have a free hand in foreign marketing. Since SMEs must rely on either government loans or internally generated funds for financing their investments, governments have an interest in the profitability of SMEs. There is also a mutual interest between the SME and the government to avoid losses that must be covered from the government budget, and as owner the government has an important stake in the profitability of the enterprise. Thus there are usually strong forces within the government favoring profit maximization of an SME and working against requiring the SME to serve special social interests, such as regional development and employment subsidization.[1]

Production and Marketing

A common criticism of SMEs by private mining companies is that they do not reduce production (or sales) in response to a decline in world prices, but rather continue to maintain output at full capacity. For example, during the early 1980s when world copper prices declined, some American copper companies cut production by closing high-cost mines, but CODELCO increased its output. There are two major reasons why SMEs tend to maintain output in the face of declining prices. First, if the bulk of their output is exported, they have little or no control over world prices, while American firms producing mainly for a national market have a certain degree of control over prices in that market. Large American mining firms sell copper at producer prices that are maintained above the world price by means of a mutual cutting-back of production and sales. In other words, they operate in an oligopolistic market in which producers take into account the impact of their sales on prices. The ability to maintain national prices above world prices differs from commodity to commodity depending upon factors discussed in Chapter 3. In the absence of market power (ability to influence prices by cutting production), there is no advantage for an SME to reduce output in the face of declining prices unless prices fall below marginal costs.

The second reason is that SMEs are under considerable government pressure to maintain employment and not close down high-cost operations. This same pressure is often exerted by host governments on mining subsidiaries of MNCs. Thus while Phelps Dodge cut its output in the USA by more than 50 per cent in 1982 following a sharp decline in copper prices, SPCC (which is partly owned by Phelps Dodge) increased output.

Most large mining SMEs have been fairly successful in selling products such as copper for which there exists a competitive world market. However, in the case of commodities such as iron ore and bauxite for which open markets do not exist, SMEs often form joint ventures with foreign companies that purchase these products for use in downstream operations. In some cases they arrange barter deals with governments of socialist countries, but such arrangements are often less profitable than selling under contracts to consuming firms in the industrial countries.

Sources of Financing

Governments usually do not endow SMEs with a substantial amount of equity financing – they must rely heavily on borrowing or reinvested earnings. However, since a number of SMEs were organized to operate expropriated foreign properties for which the foreign owners were allowed only a small compensation in relation to asset value, many SMEs came into existence with a relatively large amount of equity. A substantial amount of financing for SMEs is derived from foreign borrowing, much of it in the form of international commercial bank loans carrying explicit or implicit government guarantees. It is estimated that in 1978 nearly one-third of total Third World commercial borrowings for all purposes went to government enterprises, but it is not known how much represented loans to mining firms (Gillis, Jenkins and Lessard, 1982). In addition, there have been a number of loans by the World Bank and the Inter-American Development Bank (IADB) to Third World governments for transfer to government mining firms. SMEs also borrow from domestic credit institutions, in large part from government-owned banks. Where the SMEs have been profitable, as in the case of CODELCO and CVRD, reinvested cash flow from operations has been an important source of financing. However, a large number of mining SMEs have had little earnings to reinvest.

Most SMEs have paid few dividends to the government (Gillis, Jenkins and Lessard, 1982). SMEs are usually subject to corporate profit taxes although they may not pay the same rate paid by private firms. In years of high earnings SMEs, such as CODELCO, Centromin (Peru), Ferrominera (Venezuela), and Zambian Consolidated Copper Mines Ltd, have paid taxes equal to about half of their net profits.

Illustrations of Operations of Two Mining SMEs

CODELCO
CODELCO operates four mines formerly owned by American companies: Chuquicamata and Salvador (Anaconda); El Teniente (Kennecott); and Andina (Cerro). CODELCO is the world's largest copper producer with an output of about 1 million mt. Its production costs are among the world's lowest and it is one of the few large copper companies that operated at a profit during the period 1982–4. CODELCO is well managed and maintains a substantial degree of independence from government bureaucracy. Its efficiency is indicated by its ability to reduce net operating costs by more than 50 per cent between 1974 and 1977 in the face of rising average costs in the rest of the world. This was achieved by large expenditures for plant modernization, by maintaining a high rate of capacity utilization and by a variety of economies. The latter included a reduction in energy consumption per ton of copper produced, and a 16 per cent reduction in the labor force between 1976 and 1983, accompanied by a 20 per cent increase in output. It seems clear that maximum profitability and

growth have been CODELCO's major objectives. In 1983 and 1984 it had operating profits of $528 million and $332 million, respectively. In 1983 it paid over $300 million in taxes, and paid dividends to the Chilean government of $178 million. CODELCO has benefited from low-cost foreign loans which, together with reinvested cash flow, financed $1.7 billion in new investment between 1976 and 1983 (in 1982 dollars) (Contreres, 1984, pp. 47–57). The company publishes an annual report audited by Coopers & Lybrand, and its accounting practices appear to be much the same as those of a private US company.

Unlike CVRD and other SMEs, CODELCO has not branched into production other than coppper and the byproducts of its mines, nor has it undertaken the development of new copper mines. Although CODELCO's output represents about 80 per cent of Chile's total copper production, there is another SME, Empresa Nacional de Minera (ENAMI), and several medium-size, privately owned copper companies. The Chilean government welcomes foreign investment in copper mining, and there is no evidence that CODELCO has sought to prevent private investment in the copper industry or to take over existing companies.

Zambian Consolidated Copper Mines (ZCCM)

The performance of the Zambian state copper SME, Zambian Consolidated Copper Mines (ZCCM), is in sharp contrast to that of CODELCO. In 1970 the Zambian government acquired 51 per cent of the equity in the country's copper industry, which was owned by British, South African and American interests and managed by two groups, AAC and Rhodesian Selection Trust (RST). The takeover involved a reorganization of the industry into two corporations, Nchanga Consolidated Copper Mines Ltd (NCCM) and Roan Consolidated Copper Mines Ltd (RCM). The government's equity in each was held through a government-owned holding company. The former owners became minority shareholders in the two operating companies, but maintained managerial responsibility through management contracts.[2] In 1974 the government cancelled the management contracts so that NCCM and RCM became self-managing companies with their chief executives appointed by the government. In 1982 the government merged the two companies into a single corporation, ZCCM, in which it holds a 60 per cent equity interest, with the remainder held by the former foreign owners. The minority equity holders have had little, if any, influence on management or policies of ZCCM and have received few dividends since 1974.

Zambia's copper output reached an all-time peak of 755,000 mt in 1969, representing 13.7 per cent of world copper output. Thereafter its share of world output steadily declined and by 1981 output was 568,000 mt, representing 6.8 per cent of world output. During the 1974–81 period the number of employees rose while labor productivity in terms of output per employee declined. Costs per pound of copper produced rose (in constant 1981 dollars) and average annual profits during the 1975–82 period were only 15 per cent of average annual profits during the 1971–4 period (Radetzki, 1983, Chapter 7).

ZCCM pays a net profits tax in excess of 50 per cent and, in addition, paid a tax of 8 per cent on the gross value of mineral exports during 1983, which tax is not allowed as a deduction in calculating income taxes. The company has been unprofitable in recent years, but in 1982 and 1983 invested $275 million and $250 million respectively in cost-reduction programs. Funds for these programs were largely financed by the Zambian government and by loans from international development institutions.

The Zambian national industry's poor performance after 1974 can be only partially attributed to nationalization and the departure of the former owners. In addition to low copper prices (which affected all copper producers), Zambia had severe transportation problems arising from closure of the border with Rhodesia; a major disaster at the Mufulira mine resulted in a substantial loss of production; and the mines have had difficulty in retaining European expatriate professional and managerial personnel upon which they are heavily dependent. Since Zambia is dependent upon exports of copper (and its byproduct cobalt) for over 90 per cent of its foreign exchange earnings, periods of low earnings have led to shortages of exchange for imports of materials, equipment and services required by the industry itself.

Despite these conditions unique to Zambia, there is considerable evidence that a portion of the poor performance of its industry may be attributed to nationalization. Unlike Chile and some other economically advanced Latin American countries, Zambia does not have a supply of domestic mining engineers and trained managers, and it will be decades before such personnel become available. The loss of expatriate personnel can be traced in considerable measure to the departure of the foreign companies and to the reduced pay and amenities afforded to expatriates following nationalization.

There is considerable evidence that ZCCM has pursued social objectives, such as maintaining or increasing employment of Zambians, at the expense of profit maximization. Most successful mines in other countries have substantially reduced the ratio of workers to output through increased mechanization with new technology, and improved operating efficiency. There was also a deliberate effort to replace expatriates with Zambian personnel who were not adequately trained. The lack of emphasis on profit maximization may reflect both the character of top management following nationalization and the considerable central government control over the industry. Recently the company has been engaged in a modernization program designed to reduce costs, but it will require several years of substantial investment before Zambian copper becomes cost competitive with more efficient copper producers.

Conclusions on State Mining Enterprises
Most large SMEs are well managed and efficiently operated. This has been especially the case where expropriated enterprises inherited a well-trained professional and managerial staff from the displaced foreign owner. In some cases managers have pursued social objectives, such as maintaining employment at the expense of productivity, but in other cases profit maximization has been the dominant goal. The productivity of some SMEs has suffered from inability to obtain foreign exchange for material supplies and plant modernization. Although cost comparisons among companies in different countries are difficult to make, strong evidence does not exist that, given the same physical conditions, production costs in state-owned mines are significantly higher than those in privately owned ones. The principal advantages of SMEs over MNCs are lower capital costs and in some cases lower taxes. Affiliates of MNCs have the advantage of drawing on the technology and professional personnel of their parent companies and they frequently have a marketing advantage. Although mining SMEs were able to obtain capital from official and private foreign sources at relatively low cost during the 1970s and 1980s, capital from these sources will be less readily available during the coming decade and may tend to impair the growth of SMEs during the next decade.

Joint Ventures with Foreign Enterprises

Over the past fifteen years Third World governments have formed a number of joint ventures with multinational mining companies. For purposes of this discussion, a joint venture is an organization, in corporate or partnership form, in which ownership is held by one or more foreign investors and a government or SME, with a substantial measure of equity ownership held by each entity. It must be distinguished from a "participation" in which the host government holds a relatively small percentage (less than 25 per cent) of the shares of a subsidiary of an MNC. We may also distinguish between a *pure* joint venture in which the initial equity capital of a mining venture is supplied jointly by the foreign investor(s) and an SME, and a joint venture in which a mining company was initially established by an MNC, but a portion of the equity of the company was subsequently acquired by the government. The latter is frequently referred to as a "partial nationalization."

Joint ventures between governments and foreign investors in Third World mining are usually motivated by political considerations on the part of one or both partners. The government wants to maintain a high degree of control, or at least the appearance of control, over the natural resources, while a foreign investor may believe a joint venture is less subject to nationalization or contract violation than a wholly foreign-owned mining subsidiary. However, both the government and the foreign investor may also have nonpolitical motivations. Joint ventures in bauxite and iron ore enable foreign aluminum and steel companies to obtain an assured source of raw materials with a smaller capital outlay and risk. An SME may seek foreign joint venture partners to assure a market for its product by negotiating long-term contracts with partners that have a substantial equity interest in the project. International loan capital is usually more readily available for a joint venture. Joint ventures involving governments are eligible for World Bank and IADB loans, provided the government guarantees the loan. Private international lenders are often more willing to finance a venture with which a well-known multinational company is affiliated. In a number of cases, joint venture projects have been financed by a combination of World Bank and private commercial bank loans. Thus a joint venture between an SME and an MNC may have cost advantages in obtaining foreign debt financing.

The success of a joint venture depends in large measure on the mutuality of interest in maximizing the profits of the enterprise. This condition is enhanced if the government partner, such as an SME, is reasonably independent of the central government bureaucracy and its management has an interest in maximizing the cash flow of the project. This is the case with several joint ventures entered into by CVRD. One such venture is Mineracao Rio do Norte S.A. (MRN), organized in 1976 to develop bauxite deposits in northeastern Brazil. The deposits were discovered by Alcan in the 1960s, but it needed partners because development – including a railroad, ore treatment and shipping facilities, and a townsite for 4,600 people created in a remote and largely uninhabited area – required a larger financial commitment than Alcan wanted to make. The major partner, CVRD, has 46 per cent of the equity; Alcan has 19 per cent; and Norwegian, Brazilian (private), Spanish, US and Dutch companies hold smaller equity shares ranging from 5 to 10 per cent. The partners provided $100 million in equity financing and external sources provided loans totaling some $400 million. The managing director of MRN is appointed by the Brazilian Minister of Finance, but there is a two-man executive committee – one man appointed by CVRD and the other by Alcan – that approves all major decisions. Alcan also has the right to name the financial and administrative managers. Initially the foreign partners purchased all the bauxite, but beginning in 1984 CVRD began to acquire some bauxite for use by another joint venture in partnership with foreign aluminum companies. CVRD also has a joint venture with a group of Japanese steel companies that provide a portion of the equity and have long-term contracts for the purchase of iron ore.

CVRD is a well-managed and successful SME that has considerable independence in using after-tax earnings for reinvestment, and its management has considerable independence from the central government. By contrast, there are a number of joint ventures between foreign investors and government agencies that have no independent existence as operating companies and have little vested interest in the profitability of the joint enterprise. In some cases the joint venture itself is not a corporation, but simply a partnership between the government and an MNC, which provides for a division of the output of a mine between the two entities and certain financial obligations undertaken by each partner. In such instances there may be no mutual interest by partners in maximizing net profits. The government may be interested in maximizing employment and promoting regional development with the funds of the joint venture rather than minimizing the economic costs of the operation. In such cases conflicts between the partners are very likely to arise.

Joint ventures resulting from a partial nationalization in which the government acquires a majority interest in a foreign subsidiary are usually not successful unless management is firmly in the hands of the foreign partner and unless the success of the venture is heavily dependent on the purchase of output by the foreign partner. Mention was made in the previous section of the nationalization of the Zambian copper industry which resulted in the foreign partners becoming minority stockholders with no role in management. However, two bauxite joint ventures – one between Kaiser Bauxite and the Jamaican government and the other between Reynolds Metals and the Jamaican government – have proved to be successful. In both cases the Jamaican government acquired 51 per cent of the mining assets of the foreign companies in 1977. The foreign partners purchase all the bauxite and are responsible for management.

One pure joint venture in which there is considerable mutuality of interest in the financial success of the partnership is the Colombia nickel mining company, Cerro Matoso S.A. (CMSA). The equity in

CMSA is held by a Colombian government enterprise (47.5 per cent); a subsidiary of Hanna Mining Company (6.0 per cent); and Billiton Metals & Ores (46.5 per cent). Hanna serves as manager and provides the technical expertise required for production of the end product, ferronickel, and development of the complex mining operation. Billiton markets the product which, unlike copper, cannot be readily sold on world markets without special relations with consumers. The SME partner is basically a paper organization in which the managers serve mainly as agent for the government. There have been some conflicts between the government and the foreign partners, but the government has a strong interest in the profitability of the enterprise in which it has invested substantial funds and, in addition, has guaranteed a World Bank loan to CMSA. Moreover, the viability of the joint venture is heavily dependent upon the managerial, technical and marketing services of the foreign partners.

Conclusions on Joint Ventures
Joint ventures provide some promise of avoiding conflicts between governments and MNCs, and may provide a mechanism for increasing foreign investment in minerals. There are, however, significant limitations. There are inevitable conflicts between governments and MNCs over the division of earnings and these are unlikely to be moderated by the existence of a joint venture. However, an independent profit-maximizing SME in partnership with an MNC may support the interest of the joint venture in matters relating to taxation, government regulation, foreign exchange controls, employment, local purchases, and social expenditures having an adverse effect on profits. It must be recognised, however, that a strong SME may have a conflict of interest with the foreign partner. For example, the SME may decide to obtain mine products for its own downstream operations at the expense of sales to the foreign partner. A conflict may also arise if the foreign partner reduces purchases from the joint venture in periods of low demand, or in favor of some other source of supply. A final conclusion is that if an SME partner has no vested interest in the profit maximization of the joint enterprise, the venture is unlikely to have any substantive advantages over traditional foreign investment.

Notes

1 For a discussion of SME behavior in relation to government policies, see Escobar (1982, Chapter 6).
2 For a discussion of the history of the Zambian copper industry, see Cunningham (1981).

Appendix Table 2.1a

Major Third World Mining Enterprises with 35 per cent or More of Equity Held by the Government, 1980

Country	Govt equity share (%)	Capacity '000 mt/yr	Other equity holders
		Bauxite[a]	
Brazil			
Mineracao Rio do Norte (MRN)	46	3,400	Alcan (24%) plus minor foreign interests
Ghana			
Ghana Bauxite Co.[b]	55	300	British Aluminum Co. (45%)
Guinea			
Compagnie Des Bauxites de Guinee	49	9,000	Halco Mining (51%)
Friguia	49	3,000	Pechiney, Noranda, British Aluminum, Alusuisse, Vereingte Aluminum Werke (West German) (51%)
Offices de Bauxites de Kindia[b]	100	2,500	
Guyana			
Guyana Mining Enterprise[b]	100	2,500	
India			
Bharat Aluminum Co.	100	400	
Indian Aluminum Co.	45	500	Alcan (55%)
Indonesia			
PT Aneka Tambang	100	1,800	
Jamaica			
Kaiser Bauxite Co.[b]	51	4,200	Kaiser Aluminum & Chemical (49%)
Jamaica Reynolds Bauxite Partners[b]	51	3,100	Reynolds Metals (49%)
		Copper[c]	
Brazil			
Brasileira do Cobre	100	30	

Country	Govt equity share (%)	Capacity '000 mt/yr	Other equity holders
Chile			
Corporacion Nacional del Cobre (CODELCO)[b]	100	890	
Empresa Nacional de Minera (ENAMI)	100	25	
India			
Hindustan Copper	100	35	
Mexico			
Mexicana de Cobre	44	180	Private Mexican investors
Minera de Cananea[b]	52	65	Anaconda (ARCO) (48%)
Peru			
Centromin[b]	100	34	
Mineroperu	100	33	
Zaire			
Gecamines[b]	100	662	
Zambia			
Zambian Consolidated Copper Mines (ZCCM)[b]	60	704	British, American and South African investors (40%)
Iron ore[a]			
Brazil			
Companhia Valle do Rio Doce (CVRD)	100	80,000 (est.)	
Chile			
Compania de Acero del Pacifico[b]	100	6,935	(1978 production)
India			
National Mineral Development Corp.	100	12,000	
Liberia			
Lamco	37	10,600	Swedish and American investors (63%)
Mauritania			
Société National Industrielle Minière[b]	100	6,336	(1978 production)

Country	Govt equity share (%)	Capacity '000 mt/yr	Other equity holders
Peru			
Empresa Minera del Peru[b]	100	4,854 (1978 production)	
Venezuela			
Ferrominera Orinoco[b]	100	15,300 (1981 production)	
		Nickel[a]	
Colombia			
Cerro Matoso S.A. (CMSA)	45	700	(1984 est.) Billiton (39%), Hanna (14%), Socal (12%)
		Phosphate[a]	
Morocco			
Office Cherifien des Phosphates	100	20,000	
Tunisia			
Compagnie des Phosphates de Gafsa	98	4,600	Public stock investors
		Tin[d]	
Bolivia			
Corporacion Minera de Bolivia (COMIBOL)[b]	100	25	
Indonesia			
PN Tambang Timah	100	28	
Malaysia			
Malaysia Mining Corp.	71	15	Private investors (29%)

[a] Ore
[b] Originally developed and wholly owned by a multinational mining company
[c] Copper metal, blister or refined
[d] Metal content of ore

Sources: Radetzki, 1983; Vernon and Levy, "State-Owned Enterprise" in Jones (1982); Labys, "Role of State Trading," in Kostecki (1982), Chapter 4; and Engineering and Mining Journal, *International Directory of Mining*, New York: McGraw-Hill, 1981.

3

Decision-Making in the Mining Industry

Introduction

Investment decisions in the mining industry reflect the objectives and business strategies of the investor outlined in Chapter 2. The investment decision on a mining project is based on certain basic principles underlying all such decisions, but mining projects require special procedures for project analysis. In this chapter we begin with a brief presentation of general procedures that apply to almost any industry when making investment decisions. This is followed by a discussion of the investment decision at each stage in the creation of a mining complex from exploration to operations.

Since mining is a high-risk industry, an investment decision must take into account not only the projected financial returns but also the risk that these returns may not be realized or may not be sufficient to warrant the investment. Risk is always greatest at the exploration stage and varies with the type of exploration – grass roots or intensive exploration of an orebody on which there is considerable knowledge – but there are risks associated with other stages as well. For example, there is the risk of cost overruns in the construction stage, and in the operating period the risk that costs will prove higher than expected or that the price for the product may be lower than anticipated. While risks cannot be avoided, there are ways of measuring their potential impacts on financial returns and of taking account of these impacts in making investment decisions.

General Principles of Investment Decision-Making

All rational investment decisions require an estimation of the amount of capital needed for creating an income-generating project; a projection of annual receipts and operating costs, including taxes;

and an assessment of risk. The amount of capital required to complete a project may be found to be too high, either in terms of the funds available to the investor or in terms of his strategy in allocating capital among alternative projects. In this event the project must either be abandoned or altered to meet the capital constraint.

Cash Flow Analysis
Cash flow analysis involves a projection of the annual cash outflow and inflow over the expected life of a project. The sum of the annual net cash flow gives the undiscounted returns from the project. However, this would not reflect the cost to the investor of putting his funds into the project rather than earning a return on those funds from another form of investment. Therefore, the annual net cash flows must be discounted at a rate equal to the minimum return the investor deems necessary to warrant the investment. The sum of the discounted annual cash flows gives the net present value (NPV) of the project. The internal rate of return (IRR) for the project may be determined by calculating the rate of discount applied to the annual cash flows that will make their sum or NPV equal to zero. These calculations are illustrated in Table 3.1.

In the hypothetical example shown in Table 3.1 cash flow is negative during the first five years which cover the exploration and the construction periods. During the operating period, years 6 through 20, the revenue is assumed to be constant, as are the operating costs for each year. Taxable income is calculated by deducting both operating costs and allowable depreciation from annual revenue. (After-tax net cash flow is given in row 8.) Assuming a 15 per cent discount rate, the NPV for the entire project is $72 million, while the IRR on the project is 31 per cent. However, these calculations take no account of risk. The discount rate used in a mine project evaluation ranges from 15 to over 30 per cent. A discount rate has several components, the first being the rate of interest on relatively riskless securities, or the opportunity cost of capital. A second component is based on risk analysis for the project (discussed below). A third is the premium for offsetting inflation in order to provide a real rate of return on the investment. Thus the nominal IRR must be increased by the number of percentage points representing the expected annual rate of inflation over the life of the project.

Taking Account of Risk
Risk assessment for a prospective investment determines the nature and degree of risk in terms of the probabilities of a variety of possible outcomes. In order to assess the overall riskiness of a project it is necessary to analyze the degree of uncertainty associated with each of

Table 3.1 *Hypothetical Example of Calculation of Net Present Value (NPV) and Internal Rate of Return (IRR) (millions of dollars; ( ) = negative cash flow)*

											Years									
	1	2	3	4	5	6	7	8	9	10	11	12	13	14	15	16	17	18	19	20
1 Exploration outlays	(10)	(10)																		
2 Development investment			(10)	(35)	(35)															
3 Revenue						100	100	100	100	100	100	100	100	100	100	100	100	100	100	100
4 Operating costs						(20)	(20)	(20)	(20)	(20)	(20)	(20)	(20)	(20)	(20)	(20)	(20)	(20)	(20)	(20)
5 Allowable depreciation (20% per year)						20	20	20	20	20										
6 Taxable income						60	60	60	60	60	80	80	80	80	80	80	80	80	80	80
7 Tax (50% of taxable income)						(30)	(30)	(30)	(30)	(30)	(40)	(40)	(40)	(40)	(40)	(40)	(40)	(40)	(40)	(40)
8 Net cash flow (3) − (4) − (7)	(10)	(10)	(10)	(35)	(35)	50	50	50	50	50	40	40	40	40	40	40	40	40	40	40

NPV at 15 per cent = $72 million

IRR = 31 per cent

Source: Raymond F. Mikesell, *Petroleum Company Operations and Agreements in the Developing Countries*, Baltimore, Md: Johns Hopkins University Press, for Resources for the Future, 1984, p. 32.

the variables used in the projection of annual net cash flows. Some of these variables, such as the cost of land or equipment for which contracts with specified prices can be obtained, are reasonably certain, but other variables, such as wages, product prices and the time required to complete construction of the project, are subject to varying degrees of uncertainty. There are, however, two types of uncertainty: (1) those regarding which information exists for assigning probabilities to each of a range of possible outcomes; and (2) those for which there is no basis for assigning probabilities. An example of the latter uncertainty might be an earthquake or a flood in an area where none has ever occurred before, or some political event, such as war or revolution. There are other variables, such as prices that have displayed patterns of fluctuations in the past, or equipment breakdowns that have tended to occur with known frequencies and time patterns in the same industry, to which probabilities of variation may be assigned on the basis of past performance. Probability assessment for each variable used in a cash flow projection requires a determination of probability distribution for each of several values for the variable. For example, there may be an 80 per cent probability that product prices will be equal to or higher than the expected level and a 20 per cent probability they will be 20 per cent or more lower than expected. Mathematical techniques can be applied to determine the probability distribution of the variation of overall returns from the project. The calculated probability coefficients can then be applied to the NPV or IRR of the project to determine the probability-weighted NPV or IRR. For example, if there is a 75 per cent likelihood that the NPV will be equal to or higher than the projected level, say, $100 million, the probability-weighted NPV will be $75 million. Likewise, if the projected IRR is 30 per cent, the probability-weighted IRR will be 22.5 per cent. However, a complete risk assessment would need to include probabilities of variation of outcomes for a range of values above and below the expected NPV or IRR as determined in the cash flow analysis.

As noted above, there are some risks to which probability coefficients cannot be assigned on the basis of an objective analysis of available information. The investor, therefore, may adjust (upward) his minimum acceptable NPV or IRR to allow for such risks. This adjustment is likely to be subjective, but will be influenced by the size and financial condition of the firm in relation to the size of the investment.

There is yet another element of risk that should be mentioned. If an investment has a 75 per cent probability of success, a 25 per cent increase in the minimum acceptable NPV or IRR can compensate for a 25 per cent probability of a loss, provided the investor is *risk neutral*. If a firm is planning to invest in 100 projects, each with a 75 per cent probability of success on which the probability-adjusted NPV is 25 per cent higher than the minimum acceptable NPV, there will be some losses but the surpluses over the expected outcome on the successful projects will tend to offset the losses. This is called the portfolio effect. In this case the investor is likely to be risk neutral. On the other hand, if a relatively small firm plans to make only one or two investments that represent a substantial portion of its assets, it is likely to be *risk averse*. In this case the investor will want greater odds. For example, he might require an expected NPV that is 50 per cent higher than his minimum acceptable NPV.

Types of Risk and Risk Management

Each stage in the development of a mine is subject to special types of risk which call for risk management, including assessment of success probability and measures for limiting risk. The highest risk is associated with grass roots exploration in which the probability of failure is high and difficult to assess except on the basis of experience with, or knowledge of, similar efforts. Even after a discovery is made and extensive exploration carried out for estimating the value and quality of the mineral reserves, there is always the chance that reserves are overestimated or that the quality in different areas of the deposit may vary in some unexpected way.

There is technical risk in a mining venture, especially in the metallurgical process. Pilot plants may work perfectly, but at full-scale operation severe difficulties may arise that require a costly redesign of the plant. A number of nickel mines have experienced such problems.

Cost overruns are almost the rule in mining due to several factors, including technical problems, delays in receiving mining equipment, shortages of trained personnel, and inept management. Some of these factors contributed to the financial troubles of the Selebi-Phikwe mine in Botswana which had a capital cost at completion several times the initial estimated cost. Failure to meet the expected mine completion date increases the interest cost on borrowed capital and may require additional borrowing at higher rates of interest.

Once in operation, a mining investment is still subject to a number of risks that can lead either to low or no profits or to an early shut-down of the mine. A rise in the price of petroleum together with a fall in the price of nickel were the principal reasons for 1980 closure of the Exmibal nickel project in Guatemala, which operated for less than three years with large losses.

Finally, there are political risks arising from unexpected changes in taxes or in environmental regulations, or, as is common in developing countries, from expropriation or violation of contracts. Political risk in Third World countries is discussed in the following chapter.

There is little objective basis for establishing probabilities for many of the risks outlined above. Moreover, unlike assigning probability coefficients for a range of product prices, variations in expected returns from some events are not symmetrical. For example, technical difficulties can only give rise to a loss, never a gain. Past experience is some guide, but every mining project is unique, as is every economic and political environment. The establishment of a minimum acceptable IRR that takes into account risks for which probabilities are unknown differs from firm to firm, but in general we have the impression that the IRRs used in most projects do not fully compensate for risk. In the case of small and medium projects, promoters may not be risking a large amount of their own capital, while the equity holders are not fully aware of the many calamities that can befall a project. In the case of projects sponsored by large mining firms, there is not only motivation on the part of management (as opposed to the stockholders) to promote the growth of the firm, but there are ways of at least limiting risk exposure. One way is to minimize equity investment and finance the bulk of the capital costs with project loans on a nonrecourse basis. Nonrecourse project loans (discussed in the following chapter) are widely used by MNCs for financing projects in developing countries. Large multinational firms are frequently well diversified; they make a number of risky investments over a period of a few years and, therefore, tend to be less risk averse than smaller companies.

Stages in the Construction of a Mine

Each stage in the construction of a mine requires a separate investment decision, even though the alternative to further investment may be the loss of all that has been invested to that point. Only future revenues and costs are relevant in deciding whether to invest in the next stage of mine development. For example, an investment in an exploration program that yielded no discovery has no value except for information and experience obtained which may be applied to another exploration program. The following paragraphs examine the use of cash flow and risk analysis for making investment decisions at different stages in mining.

The Exploration Stage
Mineral exploration conjures a storybook image of an old prospector with his pick, pan and burro. It has the aura of adventure, and even today exploration in remote areas is a rugged and exciting undertaking.

Although much exploration centers on known mining districts, many companies embark on a treasure hunt. It is the lure of finding a major bonanza – a Comstock Lode or a Kidd Creek – that often gets companies into exploration. But the odds are heavily against finding large rich ore deposits, and exploration with this as the primary objective is purely a gamble in which the probability of success is extremely low. Successful mining companies generally conduct exploration in a systematic and scientific manner. This means the investment decision at each stage must be made by calculating the probability of success on the basis of experience and estimating the probability-adjusted NPV of the revenue from the exploitation of a discovery. Moreover, careful consideration is given to the alternatives to exploration as discussed in Chapter 2.

Table 3.2 shows the stages in the overall mineral exploration process. Several basic decisions must be made before starting an exploration program. A company must decide what to look for and how to look for it. What to look for is generally decided at a high level in the corporate structure, say, the vice president for exploration. The decision is often made on the basis of market projections by a mineral economist. Often mining targets are classified on a priority basis. A classification that would be relevant in 1985 is as follows:

(1) High priority: gold, platinum
(2) Moderate priority: silver, zinc, gallium, germanium, beryllium
(3) Low priority: nickel, tungsten, chromium, cobalt, mercury, tin, titanium, manganese, bauxite
(4) Lowest priority: uranium, copper, molybdenum, iron, lead.

Some flexibility in following priorities is desirable due to the possibility of finding polymetallic deposits and because geographic considerations may modify the priorities. In addition, field geologists need some leeway to pursue targets of opportunity. Even the lowest priority commodity is worthwhile if you find a superior deposit.

There are several steps in the exploration process. The first is to conduct a literature search of the geology of the region in which exploration is to take place. Included in the literature are maps, air and satellite photos, geochemical and geophysical data, and other geological information on an area. Knowledge of the geological environment provides clues regarding the mineral deposits likely to be found in an area, including those well below the surface. Ore deposit models have been formulated that describe common charac- teristics of particular types of gold or lead, zinc, silver deposits, or massive copper/zinc sulfide deposits. Knowledge of these models provides the geologist with a guide to hidden deposits of particular

Table 3.2 The Mineral Exploration Process

Phase	Stage	Activities	Goals	Major evaluations
Exploration planning	(1) Commodity and area selection	Conduct political, economic & geological studies Investigate land tenure regulations Establish exploration bases	Select commodities & favorable areas	
Finding resources	(2) Regional reconnaissance	Geological (library, mapping, remote sensing Geophysical (airborne/ground surveys) Geochemical (stream/lake sediments, soil sampling) Legal/land (survey of ownership)	Identify anomalies by ground region	Target evaluation
	(3) Ground prospecting of targets defined by reconnaissance surveys	Broad-scale mapping Detailed surveys Soil & rock follow-up surveys, grid sampling Land acquisition	Confirm and define anomalies by ground surveys	
	(4) Ground prospecting of anomalies & targets from Stage 3	Detailed mapping & sampling, lab studies Ground & in-hole surveys Rock & core studies Negotiate exploration & development agreements	Discover mineral showings	
	(5) Discoveries	Diamond drilling to identify potential deposits Possible in-hole surveys Lithology and alteration studies Acquisition agreements signed	Identify mineral deposits	Prospect evaluation

Phase	Step	Tasks	Objective	Evaluation
Proving reserves	(6) Systematic drilling & development work	Surface & underground mapping/sampling Drilling/identify ecological problems Possible underground exploration & development Estimation of reserves/financial analysis	Prove reserves	Exploration evaluation
	(7) Evaluation of reserves in mineral deposit	Determine optimum mining/metallurgical techniques Market, environmental baseline studies	Prove a technically feasible & financially interesting project	Pre-engineering evaluation
Feasibility & decision	(8) Planning feasibility study	Engineering, environmental & market feasibility studies Financial feasibility studies	Prove economic feasibility of mining reserves	Final feasibility
	(9) Final decision	Decide if deposit should be mined now or later Decide to sell or otherwise dispose of deposit	Prove viable ore deposits	
Development to production	(10) Mine development	Planning & engineering design studies completed Preproduction stripping or shaft & underground development Construction of ore treatment plant Marketing contract signed	Commence mining	
	(11) Mining operation	Mining production to produce profitable operation Continue ore exploration to maintain reserve level	Production of mineral commodities	

Source: Robert J. Miller, "The Mineral Exploration Process," *NICOR Horizons*, Fall 1983, pp. 4-7.

types since specific types tend to be located in similar environments. Therefore a literature search enables the geologist to determine where the target mineral deposits may be located.

The next step in exploration is ground prospecting. A team of geologists using four-wheel drive vehicles, helicopters, horses, and foot transportation undertake broad-scale geologic mapping, stream, sediment and rock geochemical sampling, and ground geophysical surveys. The major techniques are listed in Table 3.2. A common practice is to examine old mines and prospects. Many of the mines established today are in the vicinity of old mines. Modern technology often makes mining profitable in areas where old mines have been depleted and abandoned.

If preliminary examination of a particular area is sufficiently encouraging to warrant more detailed exploration, a right to explore must be acquired either by staking a claim on public land or by leasing land, or by negotiating with the private landowners. In more intensive ground prospecting, the techniques described in Table 3.3 are used with the goal of obtaining sufficient information and encouragement to justify a drilling program. The procedures at this stage include trenching, detailed mapping, and sampling of underground material. As exploration has shifted more and more to testing hidden targets that can be examined only by means of drilling, it is important to determine a logical pattern for drill holes over the area being explored. A mineral discovery is based on a successful drilling

Table 3.3   *Direct Mineral Exploration Techniques*

Geologic	Geochemical	Geophysical
Broad-scale geologic mapping	Stream sediment sampling	Magnetics
Detailed geologic mapping	Rock sampling heavy mineral panned concentrates	Electromagnetics
Air photo analysis		Self-potential
LANDSAT analysis	Soil sampling	Resistivity
Plate tectonics analysis	Water sampling	Gravity
Leached outcrop analysis	Soil gas sampling	Seismic
Alteration mapping	Biogeochemical sampling	Luminescence
Ore microscopy	Geobotany	Gamma-ray spectrometry
	Statistical geochemical analysis of published data	

program that establishes the existence of reserves of sufficient volume to warrant development.

It is necessary to drill a number of holes and assay the cores or rock chips in order to determine whether an economic orebody exists, and to measure the volume of reserves. Sometimes drilling is supplemented by underground workings to collect large bulk samples for metallurgical testing. If success is indicated through continued exploration, more refined reserve calculations are made and a preliminary cash flow analysis is prepared. The cash flow analysis indicates whether a mine would be economical. However, before a feasibility study can be prepared it is necessary to undertake development drilling which is more close-spaced and permits estimation of measured (or proven) reserves. Development drilling also continues after a mine is established. Since development drilling is very expensive, companies do not prove more reserves than are necessary to keep a mine operating for a limited number of years.

The exploration costs can usually be expensed as they are incurred, but at some point costs may be capitalized for amortization against revenues. Regardless of the accounting practice employed, exploration must be regarded as an investment for producing future earnings from a mine constructed to exploit the orebody. The decision to undertake exploration requires the determination of the probability-adjusted NPV of a mine based on the discovery. Discovery risk, or the probability of success of an exploration program, differs with the stage of exploration. Grass roots exploration is exceedingly high-risk and the probability of a successful discovery may be 1 in 50. Even large mining firms would not be willing to spend more than a few hundred thousand dollars on exploration over broad areas with such a low probability of success. However, exploration in the form of aerial surveys may result in discovery of small areas that are worth spending larger sums on to learn more about minerals contained in the areas. Such expenditures are warranted if the probability of a successful discovery is improved, say to 1 in 25.

Once a promising orebody is discovered, more intensive exploration for delineating the orebody and obtaining samples for metallurgical testing entails an additional investment. The success ratio in terms of determining whether the orebody contains the grade and tonnage to make it suitable for undertaking the next stage – preparation of a feasibility study – is considerably improved, say to 1 in 10. The determination of such probabilities is obviously quite crude and inexact and often made on the basis of judgments by exploration managers with considerable experience under similar conditions. Also, since annual exploration budgets are usually limited, it may be necessary to decide which of several possible programs should be undertaken.

Determining the present value of a successful exploration program requires projecting revenues and costs for a potential mine during all stages through the operating period. While all this is not done in any detail until the feasibility study, the process must be carried on in order to obtain a range of estimated present values. Prior to the feasibility study, the suitability of an orebody for further investment in exploration may often be determined on the basis of past results of exploiting an orebody with similar characteristics and specifications.

The Feasibility Study
The feasibility study is a comprehensive analysis of a proposed venture to determine the technical and economic viability of the project and provide the basis for the investment decision. Feasibility studies are expensive, and can take up to 5 per cent of the total project cost. Therefore, only when a property looks very promising is such a study carried out.

The feasibility study is based on four general categories of data: (1) geologic data; (2) location data; (3) engineering data; and (4) market data. Geologic data cover all information gathered about a deposit, including sample assays, geologic mapping and geophysical studies. Location data include the geographical location, the availability of local labor, transportation and utilities, infrastructure requirements, political considerations, and local, regional and national taxes. Engineering data include the mining and processing methods, project schedule, equipment requirements and replacement schedule, labor requirements, capital and operating cost estimates, and environmental considerations. Market data include product price projections over the life of the project.

Once the physical and metallurgical characteristics of the deposit have been estimated, the optimal mining method for extracting the ore is determined. If the ore is located well below the surface, several types of underground mining methods are possible. If the deposit is close to the surface and covers a fairly large area, a surface mining method will be chosen. Concurrent with the selection of a mining method, metallurgical testing will indicate the type of processing that should be used. In most operations the ore is crushed and put through a concentrator; the concentrates are then delivered to a smelter. However, for some types of ores a heap-leaching method may be selected. In this case the crushed ore is sent through an agglomerator that binds the smaller-sized particles into lumps and the prepared ore is dumped on a leach pad. A sprinkler system sprays a cyanide solution on the ore for a period of weeks. The solution percolates through the ore, leaching out the metals. The leach solution is collected from the pad via ditches and piped into a solution pond.

The solution is pumped through a series of carbon columns where the metals are absorbed onto the carbon. When the carbon has been fully loaded it is taken to the desorption chamber where the metals are "desorbed" and put back into solution form. This solution flows through electrolytic cells and the metal is precipitated on a cathode. The cathode is then placed in a furnace which produces a metal that may require further refining.

The determination of the annual capacity and life of the operation depends on three interrelated factors: the size of the deposit, technical considerations and economic analysis. The size of the deposit dictates how large the operation can be. The amount of recoverable product in a small deposit would not justify the large expenditure required for a high capacity operation. Technical mining and processing considerations may also limit the range of possible production rates and thus the size of the operation. Some deposits are not amenable to high output mining methods, so in those cases operations will have to be low capacity. Certain metallurgical processes entail high fixed costs. This means that if output is too low, the fixed costs may never be recovered. Some metallurgical processes do not work well beyond a certain level of production, so there may be maximum as well as minimum production rate constraints.

Once the size of the operation has been determined, equipment selection is made and manpower requirements are calculated. Also, the entire layout of the operation has to be designed. This includes not only designing the mine and plant, but also the roads, buildings, utilities, waste dumps, and tailings disposal areas. Facilities for limiting environmental impacts must also be provided. Finally, a project schedule must be developed to include all the activities necessary to bring the mine into production and to maintain production throughout its life.

The next step in the feasibility study is estimating capital and operating costs. Capital costs are estimated for all depreciable components of the project, and a schedule of capital expenditures must be formulated. Start-up costs and initial working capital requirements must also be estimated. Annual operating costs, including labor, fuel, maintenance, utilities, royalties and overhead, are then determined.

Another part of the feasibility study is market analysis. Market analysis involves determining the expected price range for the products over the life of the project. Price estimates must be adjusted for expected inflation in order to estimate the real prices for the product.

In a feasibility study, allowance must be made for contingencies, such as construction cost overruns and errors in projecting operating costs and product prices. Contingency funds are usually set at 5 to 15 per cent of the project's capital cost.

An important part of any cash flow analysis is determining the impact of taxes. In the USA mining is subject to a variety of industry-specific laws. These include special treatment for exploration and development expenditures, and the depletion allowance.

The ultimate purpose of collecting and analyzing the information outlined above is the preparation of a full cash flow and risk analysis. It is on the basis of this analysis that a decision to construct a mine or make a substantial addition to a mine complex is made. The firm's management must decide whether the probability-adjusted NPV or IRR for the project is sufficient to justify the investment.

The Construction Stage
Once the feasibility study is completed and a decision to proceed made, it is normal for the mining enterprise to hire a firm to do detailed engineering design, procurement, and mine plant and infra-structure construction. The best-known US engineering and con-struction firms are Bechtel Civil & Minerals, Inc., Fluor Mining & Metals, Morrison-Knudsen, and Ralph M. Parsons Company. There are also Canadian, Japanese and European firms that perform these functions.

The first function of the engineering and construction firm is to prepare detailed designs for each of the items in the project outlined in the feasibility study. In addition to mine and milling facilities, designs must be prepared for buildings, roads, specialized equipment, power plants, water and electrical utilities and mill waste disposal. Considerable time and expense is required for detailed designs. For a small gold mining project in Nevada involving an investment of $12 to $20 million, design time may run to six months at a cost of $150,000 to $250,000. Large copper mines involving investments of $100 million or more may require as long as two years and several million dollars for the design work alone.

Construction of a large project may take several years. In the case of a lead-zinc project being built in Alaska north of the Arctic Circle, it was necessary to construct a concentrate storage and loading facility at a coastal site and a 60-mile road to the mine. For environmental and sociological reasons, the townsite could not be built near the project. Instead there will be housing facilities on the site and work crews will be flown to their homes in Alaskan communities on a two-week rotating basis. The project was initiated in 1981 and will not be brought into full production before 1990.

Following completion of construction there is a start-up period of three months to a year before full production takes place. During this period equipment must be tested and operating personnel trained. Also during this period engineering errors must be corrected and the various elements of the project coordinated.

There are serious risks associated with constructing a mine and metallurgical facility, and problems at this stage are frequently responsible for large losses to the investors or even termination of the project (see Chapter 5). Cost overruns may amount to as much as 50 per cent of the estimated construction costs. In such cases the equity investors must either increase their contributions or make loans to the project, the interest and principal repayments for which are subordinated to the debt service on external loans. This may mean the IRR on the investors' equity declines to well below the expected level. Another contingency is a delay in completing the project. Even though the project begins production on schedule, technical difficulties may delay the time required for achieving production at design capacity. This, of course, reduces revenue for meeting debt service payments and may mean the equity investors will be required to finance a portion of those payments. Technical failures in vital sections of the plant may require redesigning and rebuilding, thereby adding to the total construction cost.

The Operating Stage

During the operating period the program established by the feasibility study, including the mining plan and production rate, is carried out. However, as mining progresses information about the deposit increases, product prices change, and there are variances from projected costs that may dictate modification of the initial mining strategy. For example, the initial strategy may call for the simultaneous mining of both higher and lower grade ores, but if the product price should decline it might be economical to mine only the higher grade ore leaving the lower grade in place. A decline in price from the predicted level may call for a reduction in output. If the decline in price were regarded as only temporary, all or a portion of the output might be stockpiled. On the other hand, if the price were higher than projected, the operator might increase output. If the higher price were expected to remain, an expansion of the mill and acquisition of additional mining equipment might be justified.

Variances between estimated and actual ore parameters may also lead the mine operator to vary the mining plan from that established in the feasibility study. A rise in the price of fuel might lead to the institution of a number of fuel-saving practices or the use of alternative power sources. An analysis of actual operating costs may lead to a change in technology for certain operations or new cost-saving technology may be introduced.

Investment decisions must continually be made during the operating period. The output of a mine can usually be expanded with new investment, or new technology may become available that reduces costs. New environmental regulations may be imposed by the government that require, say, a new smelter or changes in the plant that reduce pollution but increase operating costs. In some cases the alternative may be either making a large additional investment or shutting down entirely. US mining companies have been faced with these alternatives when meeting EPA requirements. Again, the investment decision must be made on the basis of cash flow analysis adjusted for risk.

Investment decisions must also be made by firms contemplating additional processing at the mine, such as building a smelter or erecting a power plant. In such cases the NPV of the new facility is calculated on the basis of the cost saving derived from smelting at the mine as contrasted with paying fees to a smelting firm, or generating power as contrasted with purchasing power from another source.

Marketing and Pricing Strategies

The markets for primary mineral products differ from those for manufactures in that mineral products are less differentiated and their prices are subject to frequent changes dictated by the operation of competitive markets. In addition, primary products are sold to fabricators and manufacturers rather than to ultimate consumers. Although there are differences in quality among refined metals and some producers use brand names for their products, there is a high degree of substitutability among products with the same specifications. Most sales are made on the basis of contracts between producers and consumers, but refined products can always be bought or sold on open markets. In addition to sales of refined products, there are sales of metal concentrates, nearly all of which are made under contract. Prices of concentrates follow those for refined metals. Mining firms producing only concentrates may deliver them to custom smelters and refineries for processing; processing fees are charged and the refined products returned to the producer for sale. Alternatively, producers may negotiate contracts for the sale of concentrates to firms with smelting facilities, or they may be sold through inter-mediaries in the merchant market.

There is a high degree of concentration in minerals production. National markets tend to be oligopolistic in the sense that large producers in industrial countries do not sell all they can produce at a competitive market price. Individual producers establish a *producer price* and negotiate contracts with consumers for various quantities and delivery periods at this price. Producer prices for metal tend to be more or less uniform in the national market. This uniformity is achieved through price leadership (as is the case for copper and zinc in the USA) or members of a producers association may set a uniform producer price (as was the case for Canadian nickel before 1983). Producers compete for contracts in part on the basis of quality and in part on the basis of contract terms. In periods of overproduction, individual producers may offer discounts from the producer price when competing for contracts. Producer prices frequently differ from prices on commodity exchanges. In periods of strong demand, producer prices may be lower than those on the commodity exchanges since producers prefer to maintain price stability. On the other hand, in periods of low demand producer prices may be higher than open market prices. Consumers are willing to pay higher prices than those on the free market to assure uniform quality, or to be sure of fulfilling their requirements at a relatively stable price when free market prices rise in response to heavy demand. Although producers may change their prices at any time during the period of the contract, they tend to do so infrequently, while free market prices fluctuate daily.

The ability of a group of producers to maintain a producer price in the face of declining demand depends upon the portion of total output they control and the willingness of individual producers to cut back production or sales to an amount that can be sold at the producer price. If individual producers attempt to maintain sales in the face of a decline in demand by offering discounts, the effective price may fall to quite low levels because of the inelasticity of demand for the product. A producer price is often maintained in the face of lower prices on commodity exchanges or other open markets because supplies going to these markets are small. A significant shift of buyers from producers to the open market would quickly raise the open market price. The producer price of nickel was maintained at $3.20 to $3.40 per pound during the 1981–3 recession despite the low demand for nickel while the free market price averaged about $2.25 per pound. On the other hand, the US producer price for copper declined from 101 cents per pound (annual average) in 1980 to 67 cents in 1984, which more or less paralleled the decline in the LME price from 101 cents per pound in 1980 to 64 cents in 1984. This occurred despite a 25 per cent cutback in US copper production between 1980 and 1983. Copper production outside the USA actually rose by 5 per cent over the same period. Copper producers in developing countries sell at the LME price and generally do not reduce output in response to a decline in price.

Producers of copper and other metals are also in competition with secondary producers of refined metals that use scrap rather than mine products as their raw material. Since primary producers combine scrap with mine products to produce refined metals and alloys, a decrease in their output reduces the price of scrap, which in turn lowers the input costs of secondary producers. Thus between 1981 and 1983, US secondary refined copper production was maintained, while primary production (mainly from mine products) declined sharply.

Output and price strategies may depend upon producers' costs. If the price at which a product can be sold is less than direct costs for labor, fuel, etc., the proper strategy may be to close the mine. However, a producer operating several mines may shut down only his high-cost mines and continue to produce from other mines. Metal producers desiring to reduce sales in response to declining demand also have the alternative of increasing inventories rather than decreasing output. Holding inventories is a costly but useful strategy, provided the producer believes demand will pick up in a year or two. Thus in the face of declining demand, a producer can choose among alternative strategies: (1) maintaining output and selling at a lower price; (2) reducing output; or (3) increasing inventories. Another strategy is to maintain a producer price to customers buying under contract and to sell excess output in the open market. The success of this strategy depends upon how many other producers follow the same practice since if all producers sell excess output at the open market price a decline in that price will tend to increase the differential between the producer price and the open market price to the point at which buyers will obtain their requirements from the open market.

When there are no significant barriers to imports, national metal producers may be unable to maintain producer prices at levels substantially above world prices without inducing an increased flow of imports. For example, in 1982 and 1983 the average differential between the US producer price for copper and the LME price was 5.7 cents, despite an increase in copper imports of nearly 70 per cent in 1983 over the preceding year. US producers might have sold more copper if producer prices had declined to the level of the LME price. US producers' pricing strategy was undoubtedly based on a determination of what share of the domestic market could be retained with a 5.7 cents differential between the producer price and the world market price.

Pricing strategy may also take account of the effects of the price of a metal on the demand for substitutes. During the 1964–74 period, the US producer price was in most years well below the average LME price, sometimes by more than 20 cents per pound. One reason for this was the desire of US copper producers to remain competitive with aluminum, which can be substituted for copper in many uses.

Where world producers have a substantial degree of market control because they supply a large share of the world market, in periods of low demand producer prices can be maintained well above free market prices. This has been true at times for nickel and cobalt. Where producers control the vast bulk of the output, they can ignore fluctuations in free market prices. Supplies for the free market come from small producers or inventories and are limited in volume. Large consumers must be assured of obtaining their requirements and cannot depend on the free market to supply the quantities and qualities needed.

Decisions on Financing Large Mining Projects

Large mining projects have become so costly that few companies can undertake financing with their own capital. They must either be financed with borrowed funds, or partners must be found to construct and operate the project as a joint venture, or some combination of both. Even if a company decides to finance a project with borrowed funds, it must assume a liability to the creditors that may be too large in relation to its assets. Therefore, in recent years many mines have been developed as joint ventures involving two or more private mining companies, or one or more multinationals and a government enterprise. Several examples are given elsewhere in this book. There are a number of large projects that have reached the investment financing stage, only to be delayed by the inability to find a suitable joint venture partner or partners. Examples include RTZ's Cerro Colorado copper mine project in Panama and St Joe Minerals' Pachon copper project in Argentina.

There are several important conditions that must be met to assure a successful joint venture. Each partner must be financially able to make the agreed equity contribution to construct the project. Since liabilities to creditors are generally shared, each partner must be counted on to assume his proportionate share; if one partner defaults the other must assume his share. A second condition is a common objective for the joint venture. For example, if the major objective of one partner is to acquire the output of the project at the lowest cost, while the objective of the other partner is to maximize the earnings of the joint venture, there could be a conflict of interest unless the terms and conditions of sales are specified in the agreement. A third condition is that none of the partners be in direct competition with the joint venture, or produce materials and equipment required by the joint venture. A fourth condition is that one partner must be fully responsible for management, or that management be independent of the partners. Managerial responsibility cannot be divided. Ideally, the partners should choose the managing director from outside either partner and he should be given free rein in selecting all personnel. In some cases the partner selected as manager has loaded the joint venture with top personnel from his own company and this has led to friction. Finally, major decisions should be made on the basis of unanimous consent of the partners, rather than on the basis of proportion of equity held in the partnership.

These conditions are much more difficult to achieve when one of the partners is a government or a government enterprise. Governments are likely to have social, employment and other development objectives that may conflict with maximizing the profits of the joint venture. In some cases the government partner has demanded priority to purchase the product of the joint venture, while the principal objective of the private partner has been to supply its own raw material requirements.

Private firms may have no alternative but to accept a government or SME as major partner in the joint venture. This inevitably increases the risk to be evaluated in a decision to invest. However, as indicated in Chapter 2, there are certain advantages in having a government or government enterprise as a partner.

How Successful Are Mine Projects?

Although large fortunes have been made in mining and there have been large accumulations of capital starting with small investments, historically the ratio of successful projects to total projects has been low. The reasons are varied, but in most cases the failures have reflected improper cash flow and risk analysis. A recent study of the outcome of a number of mine projects initiated over the past fifteen years strongly supports this conclusion. The following scale was used for rating the outcomes of the projects: (1) operations with "excellent success" were those with an IRR on the investment greater than 15 per cent; (2) "successful operations" were those for which revenues exceeded operating costs; and (3) "failures" were those operations where revenues were less than costs. Of the mines in the survey, 20 per cent had "excellent success;" 44 per cent were "successful;" and 36 per cent were "failures." Thus, only one in five projects met or exceeded investor expectations. Moreover, when the revenues of all operations were totaled, they were less than the combined operating costs.

Table 3.4 *Sources of Problems in Unsuccessful Mining Ventures*

Problem category	Percentage of mines with problems
Ore reserves	23
Construction sequence and cost	29
Mine plan	19
Milling	36
Processing	42
Operation management	23
Market analysis	33

Source: Whitney & Whitney, Inc., Reno, Nev.

Table 3.4 lists the categories of problems incurred and the percentage of mines in the survey that experienced problems leading to failure or unsatisfactory financial results. In many cases more than one problem contributed to poor performance. It will be observed that 23 per cent of the mines experienced errors in estimating reserves; 29 per cent suffered delays and cost overruns in construction; 19 per cent had difficulties with the mine plan; 36 per cent had technical failures in milling operations; 42 per cent had failures in the metallurgical plant; 23 per cent suffered from poor management; and 33 per cent had problems resulting from a failure to anticipate market prices for products.

While no amount of planning or control could have foreseen or overcome all the difficulties encountered by these mines, many would have benefited by better planning and control. Companies most conscientious about planning and risk analysis have been the most successful and this generalization is true regardless of size. Successful companies generally apply cash flow and risk analysis beginning in the early stages of mine development.

4

Special Issues in Foreign Investment Decisions

Introduction

The basic methodologies employed for project evaluation involving investments in foreign countries are the same as those employed for domestic investments, and the technical and commercial risks are similar. However, the political risks and risks relating to the economic environment in a developing country may be quite different from those in a developed country, and these risks must be factored into the project evaluation. This is not to say, of course, that there are no political and economic environment risks associated with mine investments in developed countries. There have been expropriations of mines in developed countries (most recently in France); there have been drastic changes in tax arrangements and foreign exchange regulations; there have been periods of high inflation and overvalued exchange rates; and there have been costly changes in environmental and health regulations. Nevertheless, risk perception by investors in their own country differs from that in a foreign country, particularly with respect to investment in a developing country. Radical changes in government structure and objectives are less likely to occur in developed countries; their judicial systems provide better protection from arbitrary and discriminatory governmental action; economic disequilibrium is less severe (e.g. triple digit inflation is highly unlikely); and investors are better able to anticipate changes in economic policies.

A high proportion of foreign investment in the mining industry by American, Western European and Japanese firms takes place in other developed countries, especially in Australia, Canada and South Africa. There is also substantial mining investment by firms of other developed countries in the USA. Although such investment is frequently subject to government screening (as in the case of foreign investments in Australia and Canada), the political risks do not differ greatly among developed countries or between domestic and foreign investments in those countries.

Foreign investments in large resource projects in developing countries usually require comprehensive agreements with governments, while such agreements are normally not required for investments in developed countries, except possibly when exploring and developing minerals on public lands. These agreements often provide for special conditions relating to taxation or other fiscal obligations, the amount and nature of investment expenditures, and the period of productive operations. For these reasons the mining agreement is an important element in the decision to invest.

Foreign investments in developing countries often take special organizational forms, such as joint ventures with government enterprises, government equity participation in the operating subsidiary, and revenue or production sharing arrangements. These organizational structures involve differential sharing of risk and management between the foreign investor and the host government that affects risk and estimation of cash flow. Special forms of external debt financing may be used to minimize risk and achieve maximum leverage. These arrangements are discussed in this chapter.

The Political and Economic Environment in Developing Countries

The constitutions of most developing countries provide that minerals in the subsoil belong to the state. Therefore, investors have no property rights and must conduct mining under a contract that allows the investor to produce and sell the minerals for a stipulated period of time and under specified conditions. The mining investor may spend hundreds of millions of dollars in developing a mine and constructing processing plants and infrastructure associated with the mine, but the value of the investment is wholly dependent on the validity of the contract. In developed countries contracts with governments have substantial protection from unilateral alteration by the government through a judicial system that is independent of the legislative and administrative branches of government. This is not true in most developing countries where the judiciary tends to support the government in disputes with private entities. Moreover, there are two legal principles held by the government and the judiciary of a number of developing countries that impair security of government contracts. One principle, referred to as *rebus sic stantibus*, is that a contract may be altered by the government whenever conditions change (or are alleged to have changed) from those existing at the time the contract was negotiated. This principle is at odds with the legal principle of sanctity of contract, *pacta sund servunda*, that constitutes the basis for contract law in developed countries. The second legal principle adopted by virtually all Latin American countries is that disputes arising over contracts between the government and a foreign investor must be settled by national courts and are not subject to international arbitration. Some contracts explicitly forbid an investor from calling on his own government for assistance in dealing with disputes with the host government. Adherence to these principles explains why only three Latin American countries (El Salvador, Guyana and Paraguay) are members of the World Bank-sponsored International Centre for Settlement of Investment Disputes (ICSID). On the other hand, a number of Asian and African developing countries are members of ICSID and some of their foreign investment contracts provide for the submission of disputes to ICSID for arbitration.

The existence of these legal conditions often means that mining contracts constitute little more than the basis for continual negotiation with the government over terms and conditions under which the investor may operate. While company lawyers spend a great deal of time negotiating the details of contracts, their basic validity depends upon the incentive and will of the government to act in a reasonable manner and to promote the mutual interests of the investor and the government during the life of the contract. Political instability in the country and the politicization of relations between government officials and the mining company may lead to continual disputes and erosion of contract terms, often accompanied by a disruption of output or the inability of the mining firm to operate at maximum efficiency. Moreover, the overthrow of an existing government, either by election or military coup, is often accompanied by the violation of contract terms in the name of a "reform" of the mining industry by the new government. Since most mines require twenty to twenty-five years from the time of the initial investment before a reasonable rate of return may be realized, there is a high probability that the firm will have to deal with several quite different political regimes during the life of the contract. There are only a handful of developing countries that have a record of relative government stability over the past two decades, and each change in government has usually brought about a demand for change in mining contracts.

Severe economic instability in the host country often creates serious problems for the foreign investor. High levels of inflation usually mean that local costs rise faster than the decline in the official value of the domestic currency in terms of dollars or other international currencies. When foreign-owned mining companies exchange their international currency earnings for domestic currency at the official exchange rate, they often experience a real loss. Prices of the products sold in the domestic economy may be held down by price controls while costs continue to rise. The government may also limit foreign exchange expenditures or the transfer of earnings by the mining company. Finally, the ability of a mining subsidiary to obtain foreign loans may be impaired by host country defaults and a low credit standing. In recent years a number of mineral producing countries – Argentina, Brazil, Chile and Mexico – have been experiencing extreme economic instability. These developments have dampened the interest of international mining companies in investing in the countries.

Outright expropriation is less likely now than in the 1960s and early 1970s, but the terms of modern mining contracts are vulnerable to unilateral change by governments. Many modern contracts do not provide for safeguards against new laws and decrees covering changes in taxation, labor relations, foreign exchange and a variety of potential governmental regulations. Petroleum companies are often able to earn enough from the discovery of large oil fields to warrant investments in risky political and economic environments. Profits in mining are too low relative to risks to make long-term investments attractive in high-risk environments. Thus we believe that poor investment climates in developing countries have been the single most important barrier to mining investment in recent years.

The Financial Structure of Foreign Mining Investments

Prior to the 1960s most foreign investments in mining took the form of equity and loan funds supplied directly by a parent company to a foreign affiliate, which might be a subsidiary corporation or a branch of the parent company. The large American, Canadian and European mining and metal processing companies that sponsored foreign mining operations in iron ore, bauxite, copper and other metals usually had sufficient cash flow to finance new investments without borrowing. They tended to finance the expansion of foreign mining projects with the reinvested earnings of foreign affiliates.

Over the past twenty-five years methods of financing new foreign mining projects have changed substantially. First, large-scale mining complexes have become very costly in relation to the capital budgets of even the largest mining companies; capital outlays ranging from several hundred million to over a billion dollars for a single project are now common. Second, the wave of government expropriations and contract violations in developing countries made it desirable to reduce risk exposure by establishing a subsidiary company that would raise the bulk of the funds for the project externally rather than having the parent company provide them directly. Third, relatively low-cost sources of external financing for foreign mine projects became available. These sources include equipment loans provided or guaranteed by government export financing agencies (such as the US Export-Import Bank or its equivalent in other developed countries) and project loans from international banking consortia at rates of interest 1 or 1.5 percentage points above the London Inter-Bank Offer Rate (LIBOR). Finally, special methods of financing have been employed to enable host governments to participate with foreign investors in the ownership of mining projects. These methods have taken the form of joint ventures or host government equity participation in the operating company. Some partnerships with foreign governments have been financed by loans from the World Bank or other international development institutions.

Foreign investments in developed countries, such as Australia and Canada, are often financed by domestic equity and loan funds, with controlling equity interest held by the foreign investor. The Australian government requires foreign investors to establish programs that provide for eventual ownership of 50 per cent of the equity by Australian citizens.

Foreign Investments Financed with Project Loans
The typical large foreign mining investment in a developing country is a domestic corporation, all or the majority equity of which is held by a multinational corporation, or by two or more such corporations in a joint venture. Frequently, 25 or 30 per cent of the capital for exploration and development is supplied by equity, with the remainder supplied by equipment sellers and private international financial institutions. The borrower is the domestic subsidiary and normally the parent company(s) does not guarantee repayment of the loans. However, in most cases the foreign equity investors are required to guarantee the completion of the mine and its ability to operate for a period at a certain percentage of capacity, and to finance cost overruns.

In most cases the product of a mining complex, whether ore, concentrate, or metal, is sold on the international market. Suppliers of loans and equipment credits look to the foreign exchange proceeds from the export of products for payment. Two devices are used to assure that creditors have first claim on export proceeds. First, the mining subsidiary must negotiate long-term contracts for the sale of products, including in some cases a contract with one of the equity holders. Prices under the contracts, however, are usually based on international prices for the product, such as quotations on commodity exchanges. The second device is an agreement between the creditors on the one hand and the operating subsidiary and the host government on the other, whereby the proceeds from exports will be paid into a special account in a bank in a foreign country (usually the USA) and debt service payments will be made from this account before the funds can be drawn for any other purpose. This second device is necessary because a host government will normally require all foreign exchange proceeds to be delivered to its central bank in exchange for local currency. Under these conditions, debt service would be at the discretion of the central bank. As an additional security, the creditors may also obtain insurance against expropriation of the subsidiary or other forms of contract violation from the US Overseas Private Investment Corporation (OPIC), or similar institutions in other developed countries. The equity holders may also obtain insurance against expropriation, currency inconvertibility, or civil disturbance that would prevent production and marketing of the product.

Despite these arrangements, creditors who make project loans are subject to certain risks that normally cannot be insured.[1] Perhaps the major risk arises from the possibility that export proceeds will not be large enough to cover both service payments on loans and the minimum revenue necessary to maintain operation of the mining project. This situation might arise due to a drastic decline in the price of the product, or to an unanticipated increase in operating costs. Such developments would constitute serious errors in projecting the net cash flow of the project at the time of the feasibility study, but commercial risks cannot ordinarily be insured. If the cash flow is insufficient to cover debt service plus operating costs, the creditors may have the choice of either taking over the mine (since they usually hold a mortgage on the property), or of agreeing to scale down the debt survice. A good example of losses to creditors as a consequence of insufficient cash flow is provided by the experience of the Selebi-Phikwe nickel/copper mine described in Chapter 5.

A second source of risk to creditors making project loans would be the inability, for technical reasons, of the equity holders to complete the project, or to provide funds necessary for meeting cost overruns. The creditors would, of course, have recourse against the equity holders, but there could be costly litigation or the sponsors might become insolvent. Usually sponsors are large multinational firms with high credit standings so this risk may be minimal. In the case of cost overruns, the creditors are often willing to provide additional financing for completing a project and some creditor agreements provide for an additional amount of such financing in the event of cost overruns. Alternatively, the creditors may give permission to the mining subsidiary to borrow additional amounts from third parties and to bring the new creditors under the original creditor agreement. However, this would increase claims against the cash flow generated by the project and also increase the first risk mentioned above.

A third type of risk would be some action by the host government that violates the terms of the agreement with both its sponsors and creditors, the adverse financial consequences of which could not be fully insured. For example, the government might impose a new tax on the operating company that would severely reduce its net operating profits available for debt service.

The greatest risk in a mining project is that borne by the initial equity investor over the period from preliminary exploration through completion of the feasibility study. These activities generally account for a substantial portion of the equity investment and, except for the possibility of tax writeoffs, the expenditures cannot be recovered unless the feasibility study is sufficiently favorable to warrant construction of the mine and to attract external debt capital for completing the project. Once satisfactory arrangements are made for construction of the mine, it is sometimes possible to obtain additional equity financing from private domestic or foreign investors, or from consumers who may contract for a portion of the output. Some mining agreements provide the government with an option to acquire a certain amount of equity either at the time construction is initiated, or, more generally, at the beginning of production. Large consumers may provide equity and/or debt financing as a means of assuring a supply of raw materials. For example, Japanese steel firms have provided such financing in connection with long-term contracts for iron ore, with repayment of the loans made by the delivery of iron ore over the life of the contract.

Joint Ventures with Foreign Governments
An increasingly popular form of foreign investment is the joint venture involving one or more foreign private investor(s) and the host government. There are a number of examples of Third World mining enterprises which were developed as wholly foreign-owned enterprises in the past, but which subsequently became partially owned (50 per cent or more) by the host government. This occurred in the Kaiser Aluminum & Chemical bauxite operation in Jamaica and in the large foreign-owned copper operations in Zambia. However, a true joint venture involves a substantial partnership interest by the government early in the project, with exploration and development risks shared by it and the foreign enterprise(s). Examples of such joint ventures include Mineracao Rio do Norte (MRN), a Brazilian producer of bauxite, which is owned 46 per cent

by the CVRD, with the remaining equity held by four foreign aluminum companies; Cerro Matoso, a nickel mining and smelting project, which is owned 45 per cent by a Colombian SME, with the remaining equity held by Hanna Mining Company, Standard Oil of California and Billiton; and a joint venture between a Colombian government enterprise and a subsidiary of Exxon for the development of the Cerrejon coal deposit in northeastern Colombia. In the coal joint venture each party participates 50 per cent in the investment and is entitled to 50 per cent of the coal produced.[2]

Joint ventures involving a government enterprise are usually controlled by a joint management committee of representatives from the partners, but the managing director of the enterprise may be an employee of one of the partners. Such joint ventures are often heavily financed by external capital, which may include a loan from the World Bank. For example, the Colombian nickel project mentioned above received an $80 million loan from the World Bank; a line of credit ranging from $94 to $120 million from an international banking consortium headed by Chase Manhattan Bank; and a loan package of about $26 million from the Export-Import Bank and a group of private US banks. The creditor agreement covering these loans includes a completion guarantee and the equity partners are responsible for cost overruns. Hanna Mining provides management and various technical services under a technical service agreement with the joint venture, Cerro Matoso.

The joint venture arrangement has been regarded as providing the foreign investor with greater security against contract violations by the government than in the case of investments wholly owned and controlled by foreign investors. The government and the public tend to view the joint venture as a state enterprise, even though the state may not be the majority equity holder. Joint ventures are frequently used in projects where one or more of the foreign partners contracts for a substantial portion of the output. This is the case with Brazilian joint ventures in iron ore and aluminum; and Billiton, the largest equity holder in Cerro Matoso, has contracted for a large portion of the ferro-nickel output of that project. Thus, a dispute among the partners might jeopardize the marketing of output. For these reasons it has been suggested that joint ventures might provide a means whereby developing countries could attract the financial, technical and managerial resources of multinational corporations while at the same time satisfying the preference of Third World countries for development of their mineral resources by the government.

The joint venture is especially suitable for investments in which the foreign partner is the major consumer of mine output. The foreign partner provides an assured market for the product and the joint venture is not in competition on the world market for the same product produced elsewhere by the foreign partner. The foreign investor has an assured source of raw materials for its integrated operation without having to assume the full cost and risk of the investment. In addition, the foreign investor is afforded protection against expropriation or contract violation since the domestic partner may not be able to find a ready market for the product if the foreign investor withdraws. If the SME partner is reasonably free from government interference or domination, the partners have a mutual interest in minimizing production costs and maximizing after-tax revenues of the joint venture.

Commodities such as bauxite and iron ore are particularly good candidates for joint ventures because they constitute raw materials for large integrated aluminum and steel companies. There are no organized markets for these commodities, as in the case with copper, and the integrated aluminum and steel companies prefer controlling output at every stage from iron ore through fabrication. In Brazil large integrated aluminum companies have invested in alumina and aluminum joint ventures as well as in bauxite; much of the aluminum metal produced is acquired by the international company partners.

For partners in joint ventures with SMEs to avoid conflicts of interest, it is essential that they not be in competition with one another for sales of the joint product and that they have a mutual interest in maximizing net profits. Such mutual interest requires that managers of both the SME and the joint venture maintain a strong loyalty to their companies and are not employees of the government charged with a responsibility to carry out central government policy. SMEs differ considerably in their degree of independence from the central government. Brazil's CVRD appears to be fairly independent and the managers have a vested interest in the success of both CVRD and the joint ventures in which CVRD is a party. This is not the case, however, with the Colombian SME, Carbones de Colombia (CARBOCOL), which is in a joint venture with an Exxon subsidiary, International Colombia Resources Corporation (INTERCOR), in the operation of the Cerrejon coal project. The CARBOCOL representatives on the joint venture policy board have taken positions that favor government tax revenue at the expense of joint venture and foreign partner revenue. Also, the revenue of CARBOCOL from the joint venture is not limited to a share of the net profits, but a portion is derived from royalties paid by the joint venture to CARBOCOL. In addition, the coal produced is divided 50–50 between the two partners so that the partners are in competition with one another in selling the coal on the world market

(Mikesell, 1983, Chapter IX). As a consequence of these conditions, there have been serious conflicts between INTERCOR and CARBOCOL in the management of the joint venture.

Tax Arrangements in Developing Countries

There are a variety of tax and other revenue-raising arrangements applied to foreign mining investments by governments of developing countries, but the importance of various types of arrangements differs considerably among countries. There are three general types of arrangements: (1) taxes not related to either the volume of production or to net profits, such as surface taxes and import duties; (2) taxes related to the value or volume of output, such as royalties and export taxes; and (3) taxes based on net profits. Taxes on net profits vary greatly with the regulations for depreciation allowances and other factors affecting calculation of taxable income. Unlike petroleum investments for which production-sharing arrangements, initial bonus payments and high royalty and export duties are common, developing countries tend to rely more heavily on income taxes for revenues from mining investments. This reduces the risk to the mining investor compared to a petroleum investor who may be required to pay a large bonus in order to secure a contract, plus high royalties on production – all before any net profit is realized. If, in addition, there are liberal depreciation and amortization allowances, the mining investor may pay little in taxes before he recovers his investment.

Prior to the Second World War, royalties based on output (volume or value) were the principal form of government taxes on mining, but this changed with the general introduction of income taxes. In some cases there was also a payment to the government for the concession, either in the form of a lump sum or an annual payment. Although such payments are common in petroleum, they are rare in mining, in part because it is generally more difficult to determine the value of a mining concession until a substantial amount of intensive exploration has been undertaken and a feasibility study prepared. Heavy reliance on royalties has disadvantages for both the government and the investor. Although royalties do provide income for the host government when production begins, they may lead the investor to mine only the higher grade ores in a deposit since it would not pay to mine ore at an operating profit less than the amount of the royalty. Royalties increase current costs and reduce before-tax profits, while a tax system based on a percentage of profits does not reduce before-tax profits. In addition, if the home country of the parent company allows income taxes paid abroad to be credited against the corporate income tax liabilities of the home country, corporate taxation in the host country is preferred over royalties which cannot be so credited.

Rates of depreciation on mining equipment and capital consumption allowances on other expenditures involved in the creation of a mine have very important effects on both risk and the calculated IRR on an investment.[3] The most desirable arrangement for the investor is one in which all capital outlays are deducted in calculating taxable income before any tax on net profits is assessed. However, this delays the receipt of income taxes by the host government until the investor has recovered all his capital investment. Usually, depreciation and capital consumption allowances represent a compromise between full capital recovery before profits taxes are assessed, and straight-line depreciation for equipment and deduction of other capital expenditures on an annual basis over the expected life of the mine or concession contract. Most large foreign mining investments are financed by borrowed capital with debt retirement over a period of six to eight years after initiation of production. Therefore, the investor as well as the creditors will demand a system of capital recovery allowances that gives the investor sufficient cash flow to make debt service payments over the life of the loan plus a reasonable return on equity.

Taxing Economic Rent

When the government is the owner of the minerals and makes available the mining rights to a private company, the government's objective may be to extract the maximum revenue possible consistent with the continued development of mineral resources. This is almost always true with developing countries when taxing foreign investors, but it may not be the objective of governments of industrial countries. Economic rent for resources is the maximum amount of revenue that can be taxed on a continuous basis. It is the surplus of revenue over the full economic costs of production, including a return to the investor sufficient to attract the investment. Any attempt to tax more than the economic rent must eventually reduce output since revenues must in the long run cover payments to the factors of production, including capital at risk. In formulating a tax regime that will tax all or a majority of the economic rent of a given resource, the government has the problem of not knowing the revenue outcome of the investment, or what amount of net profit will be necessary to attract the potential investor.

One of the difficulties in devising a tax system that will capture an amount of revenue approximately equal to the economic rent of a resource project is that different projects will have different probability distributions for net revenues. For example, a project may have a relatively high probability of resulting in a loss of the entire investment, but a low probability of becoming a bonanza yielding ten times the normal return on an investment. Without an opportunity to earn a very high return, no investor would be interested in the project. In order to capture no more than the economic rent, the tax system would need to take account of the range of probabilities of varying outcomes and perhaps include a factor for risk aversion. It is highly unlikely that a government would adopt a tax code with these features and even less likely that it could implement such a system in a manner that would be satisfactory to prospective investors.

In addition to taxing economic rent, tax systems should be compatible with maximizing net revenues from the project. This is unlikely to be the case for any tax based on output, such as a royalty or an export tax, since such taxes increase the marginal cost of production. Hence, the investor will not increase output beyond the point at which marginal cost, including the unit tax, exceeds marginal revenue. Two types of taxes tend to be neutral with respect to maximizing net revenues: (1) bonuses or fixed lease payments and (2) a net income tax. As was noted above, bonuses and lease payments are not suitable for hardrock minerals. The traditional tax arrangement in nonfuel minerals is a royalty plus the normal corporate income tax. Although a royalty has the disadvantage of possibly limiting production, it does provide some revenue for the host government if the venture should yield little or no profits. However, royalties plus the normal corporate income tax may be less than the economic rent of a project. Many governments, therefore, use an excess profits tax to capture economic rents higher than those represented by revenues from a net profits tax plus royalties.

Excess Profits Taxes
A number of developing countries impose both a regular income tax and an excess profits tax that applies to rates of return on book value of the investment in excess of a certain percentage. Investors generally object to excess profits taxes for the following reason. All mining investments involve risk and the decision to invest is made on the basis of a series of potential outcomes to which probability coefficients are assigned. If the estimated probability of low or no profits is 50 per cent while there is a 50 per cent probability of earning two or three times the minimum acceptable rate of return, the investment may be made. But if the imposition of a high excess profits tax precludes the possibility of earning a relatively high rate of return, the investment is unlikely to be made.

The usual form of excess profits tax is the application of a higher than normal tax on profits in excess of an amount that constitutes a threshold rate of return on accumulated capital investment. Assume the normal corporate income tax is 50 per cent, the accumulated capital investment is $100 million, the threshold rate of return is 15 per cent and the excess profits tax is 75 per cent. If taxable profits are $20 million, $15 million would be taxed at 50 per cent and $5 million at 75 per cent, so that the total tax would be $11.25 million ($7.5 million plus $3.75 million). In addition to the general objections to all excess profits taxes mentioned above, there are special objections to this form of excess profits tax. First, accounting rates of profit vary from year to year so that when an investor has low or no profits in one year he expects to be able to maintain his expected average return with higher profits in other years. If the government takes a substantial portion of the higher profits in favorable years, the investor may not be able to earn an acceptable average rate of return. Second, annual accounting profits do not reflect the discounted value of revenue on the original investment.

The difficulties with this first type of excess profits tax may be overcome by using the resource rent tax (RRT), or the DCF trigger tax.[4] Under this method, cash inflows and outflows generated by a project are summed each year to give a series of annual net flows. The net cash flows are accumulated each year and discounted at a designated rate until a cumulative positive value is attained. That value is then taxed at the RRT rate. If in any year net cash flow is negative, no tax is levied and there is no further RRT until the discounted accumulated net cash flow reaches a positive value.

The basic principle of the RRT is that in any year a tax is levied only on profits that are in excess of the amount necessary to yield a specified IRR on all capital expenditures. In the example given in Table 4.1, the IRR is 18 per cent and accumulated present value does not become positive until the eighth year of operations. At this point a resource rent tax of 60 per cent is applied to net cash receipts in that year in excess of the amount necessary to achieve the 18 per cent IRR on total capital outlays since the initial investment was made. It will be noted in Table 4.1 that during the first three years net cash receipts are all negative, representing the capital outlay for each year. The accumulated NPV for each of those years rises each year by reason of the additional capital outlays and by application of the 18 per cent discount rate (or IRR) on total capital investment. In the first year of production, net cash receipts turn positive and the (negative) accumulated NPV begins to decline. However, only in the eighth year of production (or the eleventh year of operation) does the accumulated NPV turn positive at 30. At this point a tax of 60 per cent is applied to the accumulated NPV yielding a tax of 18.

Table 4.1 *Hypothetical Example of Tax on Accumulated Net Present Value (resource rent tax)*

	Net cash receipts	A_t Accumulated net present value at 18%	$T(A_t)$ Tax on returns over 18% threshold of 60%
Expenditures in the first year of capital outlays	−130	−130	
Expenditures in the second year of capital outlays	−100	−253	
Expenditures in the third year of capital outlays	−100	−399	
Year of operation:			
1st	100	−371	
2nd	100	−338	
3rd	100	−299	
4th	100	−253	
5th	100	−199	
6th	100	−135	
7th	100	− 59	
8th	100	30	18
9th	100	—	60
10th	− 50	− 50	0
11th	100	41	25
12th	100	—	60

Note: In the second year of capital outlays, $A_t = -130 (1.18) - 100 = -253$. In the third year of capital outlays, $A_t = -253 (1.18) - 100 = -399$. In the first year of operations, $A_t = -399 (1.18) + 100 = -371$. And so on. In the eleventh year of operations, $A_t = -50 (1.18) + 100 = 41$.

Source: Based on an example by Ross Garnaut and Anthony Clunies-Ross, "Uncertainty, Risk Aversion and the Taxing of Natural Resource Projects," *Economic Journal*, June 1975, p. 287.

It should be said that the RRT is generally used in combination with a normal corporate tax deducted from net cash receipts each year it is paid. The RRT is assessed as a form of excess profits tax.

The RRT is best represented by PNG's "additional profits tax" embodied in the agreement for the Ok Tedi project. Under this agreement the investors recover all their capital at a DCF rate of 20 per cent (or 10 per cent plus the rate of interest on (US) AAA bonds, at the option of the investor); any amounts above the cash flow necessary to achieve this DCF rate are taxed at the rate of 70 per cent. The RRT has thus far not been widely used for nonfuel minerals, but has been applied in a number of petroleum agreements in developing countries (McPherson and Palmer, 1984).

Relative Advantages of Differing Types of Tax Systems

In deciding on the tax system for a mining investment, governments need to keep in mind not only maximizing their revenues, but also the timing and stability of the revenues and the risk involved in differing tax arrangements. Governments run the least risk by obtaining revenues in the form of bonus payments at the time the mining contract is negotiated. However, not only are most mining companies averse to making large bonus payments before they undertake exploration and development of a deposit, but the government may need to trade off a bonus payment against much larger revenues in the form of royalties or income taxes.[5] Alternatively, the government might ask for a large annual rental payment which the investor would need to pay as long as he was in business. This would assure revenue to the government regardless of whether the mine was profitable. However, a large rent would increase the overhead costs of the mine and would also increase the investor's risk since he would have to make the payment regardless of output or profits. The investor could always walk away from the mine if the project failed. In fact, if the mine proved to be only marginally profitable, a large rental might induce the investor to terminate operations. In any case, the government would need to sacrifice perhaps two or three times the amount of the annual rent in other forms of revenue that might have been extracted in the absence of a rental fee.

A royalty has an advantage to the government over a net profits tax since it is payable regardless of the profitability of the mine. However, a royalty, or other forms of tax based on output such as an export tax, has the disadvantage of increasing the investor's operating costs. This could reduce the life of the mine since the operator would not normally mine ore beyond the point at which marginal cost exceeds marginal revenue. If the ore grade declined as mining progressed, or if some ore was more costly to mine, the total amount mined would be less than if no royalty were imposed. The result would be a loss of government revenue as contrasted with that for a tax on net profits.

Potential tax revenue to the government will be highest in the case of taxes on net profits, but there are two disadvantages to profits taxes. First, there is the risk that net profits may be low or nonexistent; and second, depending upon the depreciation schedule, the mine may be in production for a number of years before there are net profits to tax. Thus, if the government were greatly in need of revenue, it might assess a modest annual rental, plus a royalty and an income tax. In addition, it might also employ some form of excess profits tax, but this would depend upon its bargaining position with the potential investor. Since most government revenue systems involve a combination of taxation arrangements, the outcome in terms of the time pattern of receipts and of risk depends upon the relative importance of each element in the system.

The mining laws of most developing countries do not allow much flexibility in negotiations between government officials and investors. This is unfortunate because it limits the ability of government administrators to achieve the optimum development of the country's resources. In order to attract mining firms to explore less accessible and higher risk areas, the tax terms need to differ from those applicable to mining districts where roads and other infrastructure exist and where the geology is well known. Different minerals, such as gold and copper, require different tax arrangements both to promote investment and to maximize revenue. A government can, of course, provide incentives to potential investors other than differential tax arrangements. For example, it can provide portions of the infrastructure or the mining community. However, such incentives may require government capital that might better be used for other purposes.

Trends in Foreign Investment in Mining

Foreign direct investment (FDI) in metal mining has been declining since the early 1970s. In part this has been due to slow growth in the world demand for metals, the existence of substantial world overcapacity in most metals, and low metal prices since the mid-1970s. Global data on FDI by industry groups is scant and unreliable, but since US multinational corporations account for well over half of FDI in nonfuel minerals, we may use US data as a proxy for all such investment. The book value in real terms of US FDI in mining declined substantially in all foreign countries between 1970 and 1981. The (real) book value of US FDI in mining and smelting in developing countries in 1981 was less than half the 1970 level, while the (real) book value of US FDI in the mining sector in developed countries declined by about one-third over the same period (US Department of Commerce, *Survey of Current Business*, October 1971, p. 90 and August 1982, p. 22). Much of the decline in FDI in developing countries may be attributed to expropriations in Latin America and Africa.

Another measure of US FDI in mining is capital expenditures by majority-owned foreign affiliates of US mining companies. In 1974 and 1975 these expenditures in developing countries averaged $597 million (in 1982 dollars), but over the 1978–81 period average annual capital expenditures were only $249 million. Capital expenditures in the mining sector of developed countries by majority-owned foreign affiliates of US companies declined from an annual average of $1,436 million in 1974–5 to $818 million over the 1978–81 period (in 1982 dollars) (US Department of Commerce, *Survey of Current Business*, various issues). Thus whether measured by book value or by capital expenditures, US FDI in mining declined during the 1970s and early 1980s, and this decline was greater in the developing than in the developed countries. There is also evidence that the proportion of total investment in minerals (excluding petroleum) from all sources (domestic and foreign) in non-communist countries represented by investment in developing countries has been declining since 1970. On the basis of incomplete data, it has been estimated that over 20 per cent of the total investment in minerals (excluding petroleum) in 1970 was in developing countries, but by 1980 it had declined to about 10 per cent.[6]

By the late 1970s developing country expropriations of mining investments had virtually ceased and some governments were actively promoting foreign equity investment. However, there was strong government desire for national equity participation in the mining subsidiaries of foreign investors. In addition, more recent investment contracts have provided for substantial host government control over company activities relating to employment and training of nationals, infrastructure, the environment, procurement, and project development. Host governments are not simply concerned with maximizing revenues from natural resource projects, but with making the projects contribute directly to national development objectives, including regional development of the area in which the mine is located. Some agreements go into considerable detail regarding the nature and timing of various phases of a project. In addition, government approval of annual plans during the construction period, and for further development after production is initiated, is frequently required. The contract for the Ok Tedi mine in PNG negotiated in 1976–80 provides a good example. Phase I of the project called for mining the gold cap followed after three years by Phase II, during which both copper and gold were to be mined and processed. Phase I of the operation began in May 1984 but because of low copper prices, financial and technical difficulties, the company failed to meet contractual deadlines for initiating projects for Phase II (including the copper mine, a tailings dam and a hydroelectric plant). As a consequence, in February 1985 the PNG government ordered the mine closed until the dispute between the government and the company was resolved. (Mining was resumed in late March 1985.) Under a traditional contract, a mining company would have been able to scale back development when product prices were low, but under the new type of agreement exemplified by the Ok Tedi concession, the company may not have that option (*Engineering and Mining Journal*, March 1985, pp. 27–8; and Pintz, 1984).

Multinational mining companies that made large investments in developing countries during the 1970s had to accept partnerships with the host government in planning investments and conducting operations. Investment contracts do contain arbitration clauses for dealing with disputes, but few issues ever go to arbitration. In dealings between private companies and governments, political conditions generally require that governments not lose on important issues in dispute. Hence, investors must usually depend on persuasion and diplomacy in the event of conflict rather than on legalistic protection contained in contracts.

During the 1970s mining companies sought to limit their risk in making large capital investments by financing a high proportion of the investments with nonrecourse loans. Under this arrangement, the mining company accepted the technical risks of bringing the project on stream by a specified date, while the lenders accepted the commercial and political risks of the project's generating sufficient income to meet debt service after production began. However, in a number of cases cost overruns caused by technical factors and delays in completing projects resulted in several mining companies having to meet payments on external debt from their own resources. This was the case for AMAX and AAC in their investment in the Selebi-Phikwe nickel/copper mine; for INCO in its investment in the Soroako nickel project in Indonesia; and for INCO and Hanna Mining in the Exmibal nickel project in Guatemala (Mikesell, 1983, Chapters III, VI and XI). Although losses to lenders on nonrecourse loans for large mining projects in developing countries have been rare,[7] such loans may be more difficult to obtain in the future due to the Third World debt crisis and deteriorating economic and political conditions.

A common feature of recent mining agreements has been the granting of an option to the host government to acquire a percentage of the equity in the operating subsidiary, either after the feasibility study has been approved, or after initiation of production. The granting of such options is required by law for all foreign mining ventures in Indonesia, PNG and Peru.[8] In some cases, as in the Selebi-Phikwe project, government equity participation was provided without cost to the government. Host governments view equity participation as a means of exerting greater control over foreign company policies, and in several countries the government is entitled to appoint a member to the board of directors of the operating company (Mikesell, 1983, pp. 119–20). A number of company officials have found host government participation useful in acquainting the government with problems faced by their companies, especially in matters involving government/company relations.[9] Since in nearly all cases the government's equity is acquired at book value after the risk investment has been made, such participation constitutes a fiscal device for transferring a portion of the net revenue of the project to the government.[10]

In some developing countries government policy has favored *domestication* of ownership and control of mining rather than government participation in foreign mining ventures. This has been true in Mexico where the government requires majority domestic ownership and effective control of all large mining enterprises. Thus two of Mexico's largest copper mining companies, originally established by Anaconda and ASARCO, are now majority-owned and effectively controlled by nationals.[11]

Outlook for Foreign Investment in Mining

Foreign investment in mining was at a low level during the first half of the 1980s and will remain so pending a substantial increase in prices of major metals. Only investment in gold mining has been moderately active. A number of large planned foreign investments in copper have been shelved or abandoned. These include Anaconda's Los Pelambres, Superior-Falconbridge's Quebrada Blanca, the Noranda-Chilean state enterprise joint venture, Andacollo, and the Utah-Getty Minera Utah project – all in Chile; SPCC's expansion of the Cuajone copper mine in Peru; St Joe Mineral's El Pachon mine in Argentina; and RTZ's Cerro Colorado project in Panama. Construction of several aluminum and iron ore projects is on hold pending improvement in market conditions.

An important question is where foreign investment is likely to take place when metal prices improve. A high proportion of foreign investment will continue to be attracted to Australia and Canada, although domestic firms will increasingly dominate the mining industries in these countries. Two factors will be important in determining foreign investment in developing countries – the political and economic climate and the availability of external financing. Assuming stable political-economic conditions in Brazil and Chile, a substantial amount of investment in mining and processing is likely to take place in these two mineral-rich countries. There are also large undeveloped copper and gold deposits in PNG and the investment climate in this country continues to be favorable. Nevertheless, completion of the Ok Tedi mine has been delayed by external financing problems. There are also exceptional mining opportunities in Peru, but the political outlook in that country is exceedingly poor. The same thing may be said of Zaire. Mexico and Venezuela are also mineral-rich countries, but their national policies do not favor foreign investment in natural resources.

Unfavorable political and economic conditions also affect the availability of external financing and most multinational mining companies are not prepared to finance projects costing from several hundred million to a billion dollars or more with their own resources, or to assume the full risk of borrowing such amounts for investment in developing countries. It may prove extremely difficult to arrange nonrecourse financing for high-cost projects in any developing country with a history of severe debt service problems. Commercial banks, insurance companies and other suppliers of external financing to Third World mining projects sponsored by MNCs have in the past looked primarily to the soundness of the project and to the financial standing and reputation of the MNC for security in making nonrecourse loans. However, the stability and reliability of the host government is also involved since the government is usually a party to the creditor agreement and undertakes to abide by the agreement to allocate export proceeds for debt service. A question arises, therefore, whether nonrecourse project financing will be available in countries such as Peru that have defaulted on their external debt obligations.

In our view the outlook for foreign investment in large projects in developing countries is not promising. This is not to say that foreign investments will not be made in these countries, but the size of the projects is likely to be modest and the aggregate amounts relatively small compared to mining investments in developed countries.

Notes

1 For a discussion of risk in project lending see Walter (1984, Chapter 14).
2 For a brief discussion of the Colombian nickel and coal projects see Mikesell (1984, Chapters VIII and IX).
3 The internal rate of return is that rate which equilibrates the present value of the stream of annual revenues with the present value of the costs, including all tax payments.
4 For a full description of the resource rent tax, see Garnaut and Clunies-Ross (1983).
5 The reason a bonus payment would be less than other forms of taxation may be explained as follows. Bonus payments add to the capital cost of a mine. An increase in the capital cost of, say, $1 million reduces the risk-adjusted NPV of the expected earnings by the same amount. Therefore, a potential investor may not be willing to make a bonus payment unless he is compensated by a several-fold

reduction in other forms of taxation. Assume that the minimum acceptable rate of return to the investor is 15 per cent, and the risk adjustment factor or probability that the mine will be profitable rather than a loss is 50 per cent. If the mine contract runs for twenty years, it will be necessary for the mine to earn additional net profits totaling $6.4 million over the life of the mine to offset the reduction of $1 million in risk-adjusted NPV arising from the bonus payment. Thus the investor would require a reduction in other forms of taxes totaling $6.4 million in order to justify the $1 million bonus payment. The government, would, of course, gain the interest on the million dollar bonus, while royalty payments are made over the productive life of the mine, and corporate profits taxes often are collected only in the latter stages of mine production. In addition, the government would avoid the risk of not receiving any royalty payments if the mine did not produce, or of not receiving profits taxes if the mine proved unprofitable. But the loss of other forms of revenue might be too great a price to pay for the bonus (Mikesell, forthcoming).

6   Crowson (1982, p. 52). These data include coal, but even if coal were eliminated, the proportion of total annual mineral investment in developing countries declined between 1970 and 1980.

7   The external lenders to the Selebi-Phikwe nickel/copper mine in Botswana have sustained substantial losses as a consequence of low nickel and copper prices. After the mine met the completion requirements, the external lenders did not have further recourse to the principal investors, AMAX and AAC, for meeting payments on the debt (Mikesell, 1983, Chapter III).

8   The Peruvian mining law requires that 5.5 per cent of each mining company's net income (before taxes) be paid into a labor participation account, at least half of which must be represented by "labor shares" in the mining company.

9   This view was expressed by several multinational mining company participants at the "Regional Workshop on Mineral Development in Southeast Asia," organized by the Indonesian Mining Association, Jakarta, Indonesia, September 19–21, 1983.

10 Book value or actual expenditures by the investor will be significantly lower than the probability-adjusted net present value of the equity participation for two reasons. First, capital expenditures for exploration and development of a mine over a period of several years before the government exercises its option should be compounded at a rate of interest appropriate for a risky mining investment. Second, the net present value of the investment will be substantially higher after successful exploration and construction of a mine project. If there were a market for the equity shares, the government could pay the market price which should be substantially above book value, but such a market rarely exists.

11 These mining companies are Minera de Cananea (originally established by Anaconda) and Industria Minera Mexico (which was formerly Asarco Mexicana). In the Philippines a majority of the equity of mining companies must be held by domestic investors. Malaysia has employed measures for reducing foreign equity participation in that country's tin industry.

5

Illustrative Cases of Mining Investment Decisions

The purpose of the brief case studies in this chapter is to illustrate some of the special factors that need to be taken into account in making decisions on mining investments in a Third World environment. They include: (1) negotiating satisfactory provisions in the mining agreement; (2) adopting measures for limiting risk; and (3) taking account of the risks of contract violation or expropriation by the host government. Most of the special concerns applying to investment decisions in Third World countries have to do with limiting risk, but since many risks are unavoidable, allowance should be made for them in determining the minimum acceptable rate of return to the investor. Therefore, some of these case studies examine the cash flow analysis and rate of return projections for the investment projects.

Investment in Cuajone

Initial planning for the development of the Cuajone copper orebody in southern Peru began in 1955 following the transfer of the claim owned by Cerro de Pasco and Newmont Mining to Southern Peru Copper Corporation (SPCC).[1] (SPCC is jointly owned by ASARCO, 52.3 per cent; the Marmon Group (formerly Cerro Corporation), 20.7 per cent; Phelps Dodge, 16.3 per cent; and Newmont Mining, 10.7 per cent.) SPCC held claims on three orebodies, namely, Cuajone, Toquepala and Quellevaco, all within a short distance of one another, and had negotiated an agreement with the government of Peru authorizing the development of all three under essentially the same terms. In 1955 SPCC decided to develop Toquepala first, then Cuajone, and Quellevaco last. Construction of Toquepala was initiated in 1956 and completed in 1959, but a decision to begin intensive exploration of Cuajone was not made until 1964. In 1965 the Government of Peru required renegotiation of the bilateral agreement applying to all three mines. Before the negotiations were completed on the agreement for Cuajone, a new military government espousing socialist policies came into power in October 1968 and agreement was not reached until December 1969. Prior to the final decision to construct Cuajone, which was not made by SPCC's partners until 1970, well over $100 million had already been spent.

Planning for construction of Cuajone was based on the experience with Toquepala, which had proved to be profitable during the 1960s. However, the December 1969 agreement on Cuajone was much less favorable than the agreement under which Toquepala had operated. Nevertheless, the cash flow analysis prepared in 1969 projected that the mine would be marginally profitable with all indebtedness repaid by the eighth year of production and a nominal IRR on equity investment of 8.1 per cent over the first twenty years of production. On a *real* basis (after adjustment for inflation) the IRR would be less than half this rate. However, actual capital expenditures were substantially higher than projected and unanticipated government actions following the signing of the bilateral agreement substantially reduced net earnings. Given the risky nature of the investment and the policies of the socialist government of Peru, it is somewhat surprising that SPCC made the investment at such a low projected IRR. A possible answer is that SPCC feared if it did not go ahead with Cuajone, Toquepala would be expropriated. Had the Peruvian military dictator, Juan Velasco Alvardo, not been deposed in August 1975 (the year before Cuajone was completed), it is likely that both mines would have been expropriated. The Velasco government had expropriated several large foreign investments in Peru, including Cerro de Pasco (1973) and Marcona Mining (1975).

Another factor in the decision to invest in Cuajone was a governmental decree of February 1968 that rescinded the concession agreements applying to Cuajone and Quellevaco and established new conditions for the exploitation of both orebodies. According to this decree, failure to reach new agreements and submit plans for initiating construction in 1970 would result in concessions being forfeited. It was apparent that the concession on Cuajone would be lost if a new agreement and early initiation of construction did not take place. In January 1971 the Peruvian government declared that SPCC had not complied with the law requiring submission of documentation assuring the financing for Quellevaco and, therefore, the mining concession reverted to the government. Clearly these actions were violations of the earlier agreements and must be viewed as part of the Peruvian government's program of expropriation of a number of mining properties, sugar and cotton plantations and other large foreign investments. Therefore, in 1970 SPCC faced the alternative of making a risky new investment or losing both the Toquepala mine and the Cuajone concession, perhaps for all time. Had Toquepala not been at stake and had SPCC been able to postpone development of Cuajone, the decision might well have been to delay the investment pending a more favorable political and economic climate in Peru.

SPCC has been able to meet its debt obligations on Cuajone, but dividend payments from earnings to the SPCC partners and Billiton (SPCC's joint venture partner in Cuajone) were only $141 million through 1984, or an annual average of only $18 million on total equity investment of $309 million (as of 1978). This low rate of return on equity was due in large measure to low copper prices and a more than doubling of the projected $355 million in capital costs. However, Cuajone's earnings were also impaired by certain actions of the Peruvian government following negotiation of the December 1969 bilateral agreement.

Economic Assumptions for Cash Flow Analysis
In the preparation of the 1969 cash flow analysis, the productivity of the Cuajone mine and smelter in terms of copper output was substantially underestimated; average annual production of blister over the 1977–83 period was 166,000 st as contrasted with 113,000 st projected for the first seven years of operation. This was perhaps remarkable considering the number of days lost each year to strikes.

The cash flow analysis assumed a copper price of 40 cents per pound for the first ten years of production. (Price and cost assumptions should be regarded as real prices in 1969 dollars.) In 1969 the average US producer price of copper was 48 cents per pound and the average LME price was 67 cents, both in current dollars. In 1976 when Cuajone came on stream, the average US producer price in 1969 dollars (deflated by the US price index for industrial commodities) was 43 cents while the LME price was 40 cents. Hence, in real terms, the projected 40 cents price for copper proved close to the price when Cuajone began production. Unfortunately, however, during most of the years since 1976, copper prices in 1969 dollars have been well below 40 cents per pound. Over the 1977–83 period the average LME copper price in 1969 dollars was about 30 cents per pound.

On the basis of Toquepala's experience, direct and indirect production costs for Cuajone were estimated at 13 cents per pound excluding capital costs, such as interest and depreciation. As a consequence of sharply rising fuel and labor costs, production costs were underestimated and in 1977 were about 19 cents per pound (in 1969 dollars). Since 1977, production costs have risen substantially in real terms despite considerable progress in increasing productivity at the mine.

The 1969 analysis assumed total capital cost of the project to be $355 million, or about $3,200 per st of copper. As a consequence of cost overruns, the actual capital cost was about $1.2 billion in 1977 dollars, or about $6,700 per annual ton of capacity. In 1969 dollars this represents a total cost of $620 million or $3,400 per annual ton. Thus despite the substantial rise in the total capacity costs, capital cost per annual ton of capacity did not rise proportionately because of substantial increases in annual output over the 1969 projection. Overall, Cuajone has proved to be a relatively low-cost mine and would have been quite profitable, at least before taxes, if the price of copper in 1969 dollars had averaged 40 cents per pound since the start-up of production.

Financing the Construction of Cuajone
According to the 1969 financial plan, $67 million in equity capital for Cuajone was to be derived from Toquepala's depletion reserves accumulated during 1970–4. Peruvian law required these reserves to be reinvested in Peru so it was to SPCC's advantage to find profitable outlets for Toquepala's depletion reserves not required for reinvestment. Undoubtedly the existence of these reserves constituted an important element in the decision to invest in Cuajone. This $67 million, plus $11.5 million invested in Cuajone prior to 1970, brought the planned equity contribution to $78.5 million. The remaining $276.5 million was to come from loans to cover the estimated total project cost of $355 million. The use of Toquepala's depletion reserves was part of the Cuajone bilateral agreement and SPCC officials believed the use of the depletion reserves was guaranteed by a provision in the bilateral agreement stating that "the tax system stipulated under the Cuajone agreement shall remain in force during the time period required for recovery of the investment." However, a Peruvian decree law of April 15, 1970 substantially reduced the depletion allowance available for reinvestment so that in effect most of the $67 million in equity contribution had to be provided from other sources.

In the final accounting, equity financing for Cuajone totaled $262 million from the following sources: reinvestment of SPCC's earnings from Toquepala, $190 million; new investment by SPCC partners, $45 million; and investment by Billiton as a joint venture partner in Cuajone, $27 million. The substantial increase in equity investment was required due to the more than doubling of the cost of Cuajone and the necessity of maintaining a minimum debt-equity ratio required by the external creditors. By 1977 long-term debt financing totaled $462 million and, in addition, it was necessary to borrow working capital. The creditors required a completion guarantee from SPCC's partners, together with other requirements for project financing on a nonrecourse basis. The project was completed on time and the long-term debt was scheduled to be repaid by 1986.

Owing to accelerating capital costs during the construction period, SPCC was frequently hard-pressed to raise the required additional financing. The creditors were especially concerned that the Peruvian government might nationalize the project. Adding to these pressures was the fact that under the Cuajone bilateral agreement, SPCC was committed to make minumum expenditures each year, and if they were not made the government had the right to cancel the concession and take over the project. This action might have triggered the expropriation of Toquepala, the profits of which were important for financing construction of Cuajone.

Actions by the Peruvian Government Impairing the Profitability of Cuajone

Several actions were taken by the Peruvian government following the 1969 bilateral agreement that impaired or threatened to impair Cuajone's profitability. Whether these actions constituted a violation of the bilateral agreement is subject to dispute, but had they been known in advance it is possible SPCC's decision to invest might not have been made. Perhaps the most serious of these actions was the establishment of the Mining Communities Program designed to give workers a share in both profits and ownership of all large Peruvian mining companies. In its original form the program would have given the Mining Communities (and in effect the government) a 50 per cent share of the equity in all large mining companies after a few years of profitable operations. Thus it must be viewed more as a means of nationalizing private investment in the mining industry than as a contribution to workers' welfare. The Mining Communities Program was modified in 1978 and currently each mining company is required to allocate 5 per cent of its net income (before taxes) to a labor participation account, which, according to the election of the workers, is invested in (1) labor shares in the company; (2) labor bonds; or (3) shares in other business ventures in Peru. However, the workers must elect to receive in labor shares at least one-half of the sum to which they are entitled. In addition, 0.5 per cent of each mining company's before-tax profits must be paid to the Mining Community to cover its expenses. Holders of labor shares are entitled to participate in dividends the same as other shareholders and have the right to a preferential distribution of dividends up to 5 per cent of the nominal value of the shares, provided the company's profits are enough to cover this amount. The labor bonds are to bear a fixed annual rate of return not less than the highest authorized payment by Peruvian financial companies on their deposits and have a maturity of three years. Also, each mining company is required to contribute 4 per cent of its (after-tax) profits directly to the Mining Community so that total contributions amount to 9.5 per cent of profits. Since Cuajone's profits have either been modest or nonexistent in most years, the contributions to the Mining Communities Program have not been large, except during 1979 and 1980 when they were $17 million and $13 million respectively. However, they could constitute a significant limitation on SPCC's profits in the future, especially as the number of labor shares increases.

Effective July 1, 1975, the Peruvian government decreed that the marketing of all concentrates and metals was to be taken over by the government agency, Mineroperu Comercial (Minpeco). Although the sale of copper is governed by long-term sales contracts, Minpeco receives a market commission of 1.36 per cent of the value of Cuajone exports during the capital recovery period; thereafter the commission increases to 1.5 per cent and then to 2.0 per cent six years after the investment recovery period. Minpeco performs no service for this commission. Hence, it cannot be regarded as anything but a tax that violates the 1969 Cuajone bilateral agreement which stipulated that the tax system would not be changed during the period of investment recovery.

The tax stability provision in the 1969 agreement has, however, served to protect Cuajone from the 68.5 per cent maximum corporate tax rate to which the mining industry in Peru was subject beginning in 1984, and the 20.5 per cent tax on metal exports by which Toquepala's exports were subject from 1976 to 1983. The bilateral agreement provided for an income tax on Cuajone's profits of 47.5 per cent during the investment recovery period (estimated to occur in 1986) and 54.5 per cent for six years thereafter. Thus, after 1992 Cuajone's profits will be taxed at the normal rate for all large mining companies, currently the highest in the world.

The equity investors in Cuajone would have fared better if they had put the money in a savings bank. Given the high Peruvian corporate tax rate, the mandatory contributions to the Mining Communities Program, and the large construction cost overruns, Cuajone is unlikely to yield an internal rate of return on equity investment in excess of 5 per cent over a 20-year period, even with a substantial rise in the price of copper. The development of Cuajone probably saved Toquepala from being expropriated, but the price may have been too high.

What are the lessons from the decision to develop Cuajone? First, the political and economic climate in Peru was and has continued to be too poor for a large investment requiring many years before capital repatriation and realization of an adequate return on the investment. SPCC's decision in 1985 to postpone, perhaps indefinitely, a planned expansion necessary to maintain Cuajone's output in the face of a declining ore grade, suggests the company is pessimistic about future net returns on the investment. A second lesson is that the projected IRR for investment in Cuajone on the basis of the feasibility study did not provide sufficient allowance for contingencies, such as the decline in the real price of copper and adverse governmental actions. A final lesson is that the bilateral agreement did not provide sufficient guarantees against governmental actions, such as the denial of the right to use the depletion allowance from Toquepala to finance Cuajone, and the introduction of the Mining Communities Program. Although the Peruvian military government might have violated the agreement no matter how well it was drawn from the standpoint of protecting the company, it would have been reluctant to do so because of the impact on Peru's ability to obtain external loans and direct investments.

The Selebi-Phikwe Nickel/Copper Mine in Botswana

Some investments involve a process of creeping entrapment: at each stage a decision is made to invest more money in order to avoid losing what was previously invested. This typifies the investment in the Selebi-Phikwe nickel/copper mine in Botswana in which the suppliers of both equity and loan capital incurred large losses and the government received little revenue. Yet the mine continues to be viable and shows a modest operating profit. The financial history of this project[2] illustrates some special problems in investment decision-making and provides some lessons for both equity and loan investors in mining.

Exploration of Selebi-Phikwe by a company owned by American, British and South African investors began in 1959, following the granting of a concession by the Bamangwato Tribal Authority (a British Protectorate) to Rhodesian Selection Trust Ltd (RST). Construction did not begin until the granting of the mining lease and the signing of the Master Agreement in March 1972 by the then independent Republic of Botswana. Bamangwato Concession Ltd (BCL), the operating company, is owned 15 per cent by the Botswana government and 85 per cent by Botswana RST (BRST), the equity of which is held 30 per cent by AMAX, 30 per cent by an AAC group, and the remainder by thousands of public shareholders (mostly American).

Financing of BCL was highly leveraged with about $44 million provided by equity shareholders in BRST and the remainder from loans, including $69 million from a German banking consortium, Kreditanstalt für Wiederaufbau (KFW); $18 million from the Industrial Development Corporation of South Africa (IDC); and the balance in subordinated borrowings from BRST. Phase I of the project was originally planned to cost about $143 million while Phase II was expected to be financed from BRST earnings from Phase I, or by subordinated loans supplied or guaranteed by AMAX and AAC. The Botswana government's 15 per cent share in BCL was provided without cost to the government. The government decided to provide the infrastructure for the project, including power and water facilities and transport arrangements. The infrastructure was expected to cost about $77 million, financed by a $32 million loan from the World Bank, loans from the Canadian International Development Agency, the US Agency for International Development, and loans and grants from other sources.

Despite the initial low equity commitment to a project expected to cost about $220 million, including infrastructure, the major shareholders accepted contingency obligations in excess of $300 million. These obligations included completion guarantees for Phases I and II in the loan agreements with KFW and IDC, and a guarantee of the World Bank loan.

The mining project was plagued from the start with technical difficulties. Severe metallurgical problems developed, partly as a consequence of relatively untried metallurgical and chemical processes and partly as a consequence of managerial inefficiency early in the construction period. Phase I, expected to be completed and operating at a profit in 1973, was not completed until 1977. Capital costs soared as a result of the need to rebuild or replace much of the processing plant. Planned production of sulfur had to be abandoned and BCL incurred a loss of several million dollars on a contract to deliver sulfur to a South African fertilizer firm. Operating costs were substantially higher than planned due to a rise in the cost of fuel and other inputs. By the time production reached planned capacity, the prices for both nickel and copper had substantially declined. In addition, payments on the KFW loan denominated in Deutschemarks rose in terms of dollars because of the sharp appreciation of the DM during the 1970s.

Several financial projections were made for the Selebi-Phikwe project in 1971, but it is not clear from reports available to the authors which projections were employed in the final decision to initiate production. One BCL projection estimated an equity payback in 5.1 years and a DCF on equity of 19.9 per cent (excluding all capital investment before 1971). However, a World Bank projection dated June 1971 estimated a DCF to equity investors (excluding both government equity and pre-1971 expenditures) of 12.8 per cent. The BCL projections were based on nickel and copper prices somewhat higher than prices assumed in the World Bank projection. In both cases the price projections were conservative, given the general expectations regarding metal prices for 1971. Also, the estimated capital costs of the project per annual ton were more or less in line with the costs of new nickel mines in Canada in the early 1970s. However, considering the risks represented by the obligations undertaken by the major shareholders and the record of cost overruns in most nickel projects in the past, the DCF projections do not appear sufficiently high to provide a minimum acceptable rate of return. In retrospect, the investment was a marginal one, even if all had gone according to the 1971 plan. There is evidence that the desire to avoid the loss of $44 million already invested before initiation of construction was an important factor in the decision to invest. AMAX was also influenced by its desire to provide a supply of nickel/copper matte as feed for its Port Nickel, Louisiana refinery, a contract for which was included in the Master Agreement of 1972.

By 1976 it was clear that BCL's operating profits would cover neither the interest payments on the growing indebtedness nor the scheduled principal repayments on senior debt. Also, since the Botswana government expected to obtain the bulk of its revenue from taxes on profits and from dividends on its 15 per cent equity in BCL, in the absence of net profits for BCL the government received a royalty of less than $1 million annually. Hence, the government wanted to amend the lease agreement to provide for royalties based on production. In late 1976 negotiations were initiated for restructuring the 1972 Master Agreement. Given the precarious financial position of the project, it was hoped that the senior creditors would make some concessions. The restructuring agreement was not completed until March 1978 and few concessions were made by the senior creditors. Even before this agreement was signed, it became evident that BCL's operating profit would not permit meeting the obligations under the agreement. However, the production conditions for completion of Phase I had been met and completion of Phase II was assured. This meant the major shareholders were no longer obligated to meet payments on the senior debt held by KFW and IDC and only the guarantee of the World Bank loan remained. The World Bank loan would be repaid from BCL's payments for water facilities and infrastructure required by the mine so long as the mine continued operating. Therefore, at this point AMAX and AAC could have walked away from the project and turned it over to the senior creditors, thereby limiting their eventual losses. Also, it is likely that the mine would have continued to operate since it was yielding an operating profit, even though not sufficient to meet debt service obligations.

Under the second restructuring agreement, which was not completed until April 1980, all principal payments on indebtedness to KFW, IDC and other creditors were deferred from 1980 through 1983, but not interest payments. Also, royalty payments to the government were deferred during the 4-year period, but interest was to be paid on the amounts deferred. Despite the technically efficient operation of the BCL mine during 1981 and 1982, low prices for nickel and copper reduced BCL's income well below expectations, and by early 1982 it became clear that it would be unable to meet payments on this senior indebtedness in accordance with the schedule in the second restructuring agreement. At this time the major shareholders were in a position to demand substantial concessions by the senior creditors since the alternatives were to close the mine or to turn it over to the creditors to operate or liquidate.

Under the third restructuring agreement completed in June 1982, 70 per cent of the senior indebtedness became subordinated debt of BCL, with no maturity date. In addition, payments on the remaining senior indebtedness are subject to deferral if BCL does not have sufficient cash. A substantial portion of the subordinated indebtedness held by BRST was converted into cumulative preferred shares totaling approximately $198 million. In order to keep the mine operating in 1983, the major shareholders agreed to provide an additional $18 million in the form of senior capital.

It seems likely that BCL will be able to repay the scaled-down senior indebtedness, but how much it will be able to pay on the subordinated indebtedness or cumulative preferred shares is highly problematic. It is extremely unlikely there will be any earnings on the equity shares held by AMAX, AAC and thousands of public shareholders.

Lessons from the Selebi-Phikwe Project
The experience with the Selebi-Phikwe project provides certain lessons for equity investors, creditors and host governments when structuring mining projects. For the equity investor, the project illustrates that risk is not avoided by nonrecourse loans accompanied by completion guarantees, since technical difficulties and cost overruns are exceedingly common in mining projects and such risks must be borne by the equity shareholders. Moreover, with highly leveraged projects the amount at risk to equity investors may be several times the initial equity contribution.

A second lesson is that equity investors should be prepared to minimize their losses and to evaluate additional investment solely on the basis of probable returns from such investment, and not on the basis of saving the initial investment. It is frequently impossible to achieve a satisfactory return on an additional investment designed to salvage a project that is in serious difficulty.

A third lesson is that equity investors need to factor in suitable probability of success coefficients in projecting financial returns from a project to allow for risks. Moreover, they should regard their potential liabilities in the form of guarantees as part of the equity investment in calculating their minimum acceptable returns.

For the external creditors, the project illustrates how large losses can be sustained following the completion of a project when lenders have no further recourse against equity investors. Heavy losses to creditors occurred even when the Selebi-Phikwe project was generating operating profits, but profits were too low to service the large accumulated indebtedness. Moreover, creditors should take responsibility for monitoring a project from the beginning of construction, since they usually have the greatest amount of capital at risk.

Finally, the Botswana government incurred substantial risk in relying mainly on net profits for revenue, rather than requiring a reasonable royalty on production. It should be said that the success of the Selebi-Phikwe project was not impaired by the Botswana government, which has been quite supportive of the project throughout its history. Given the importance of the project in terms of employment and development in Botswana, the government would have the most to lose by the termination of the project.

Freeport Indonesia's Ertsberg Mine

The Ertsberg mine in Indonesia's province of Irian Jaya on the island of New Guinea was the result of a risky investment in a high-grade copper deposit in one of the most isolated and difficult physical environments in the world. The mine, completed in 1972, has been a technical and financial success. The remote location of the orebody some 55 miles inland, at an elevation of 11,000 feet in a rain forest, with no roads through a vast mangrove swamp to the coast, presented formidable problems for exploration and eventual development.[3] Preliminary investigation of Ertsberg was initiated in 1960 by Freeport Sulfur (later Freeport Minerals and now Freeport-McMoRan) in collaboration with a Dutch firm that had been awarded an exploration permit for the Ertsberg area from the Dutch government. The contract to develop the mine was not awarded until 1967, following the enactment of a new foreign investment law by the Indonesian government. The previous administration under President

Sukarno had followed an anti-foreign investment policy that prevented negotiation of an agreement on the Ertsberg project, so negotiations were delayed until after the Suharto government came into power in March 1966. This was a favorable time for contract negotiations since the new government was quite anxious to attract foreign investment to develop the country's resources. It, therefore, agreed rather readily to most of Freeport's initial proposals. Subsequent Indonesian foreign investment contracts for mining were somewhat less favorable. Nevertheless, given the history of Indonesia since it became independent in 1963, the investment was regarded as subject to considerable political risk.

The Ertsberg project, which eventually cost $200 million, was a risky one for Freeport Sulfur, which had gross sales of less than $150 million in 1967 and assets of $337 million. However, there was a need for the company to diversify. Sulfur accounted for 85 per cent of Freeport's sales and the market for sulfur was weak and the future not promising. The Ertsberg deposit was estimated at 33 million tons of 2.5 per cent copper, plus 0.025 ounces of gold and 0.265 ounces of silver per short ton, but the deposit would be exhausted in about twelve years. There was evidence of nearby copper deposits, but the amount of additional ore was not determined until after the investment decision was made.

It was necessary to commit some $7.5 million before the feasibility study could be completed. The original cost estimate for the mine and concentrator plus infrastructure was $120 million. The infrastructure included a road from the coast to the mine, the mining community, a 69-mile slurry pipe for the concentrates, a huge aerial tramway linking the mine and concentrator, and port facilities for ocean transport. Considering the engineering problems faced in constructing the mine and infrastructure, a cost overrun of $80 million was perhaps modest in comparison to other projects involving fewer uncertainties and difficulties.

Initially Freeport Minerals held 87 per cent of the equity in the operating company, Freeport Indonesia (FI); 5 per cent was held by Norddeutsche Affinerie, which had a contract to purchase a portion of the annual output of the copper concentrates; another 5 per cent by Zuid-Pacific-Koper, which held the original mining exploration concession; and 3 per cent by Indonesian private interests. Later Freeport Minerals equity in FI was reduced to 81 per cent following acquisition of 8.5 per cent of the equity in FI by the Indonesian government. The initial financial plan provided for a debt-equity ratio of five to one. Assuming 40 cents per pound of copper, the original cash flow projection for Ertsberg was expected to yield a 40 per cent DCF rate on equity over the 13-year life of the project, and a

17 per cent DCF on total capital investment. The original investment would be returned in the fifth year. Although the real price of copper (in 1970 dollars) rose substantially above the 40 cents per pound level, during the late 1970s and early 1980s the real price of copper fell below 40 cents as a consequence of the rapid rise in the wholesale price index. Also by 1975, costs were more than double the 1970 estimate. While the rates of return on the Ertsberg mine have been far less than those calculated in the feasibility study, the mine has yielded a profit to the equity investors. In fact, it is one of the few copper mines initiated during the 1970s that has proved to be profitable.

Despite the high debt-equity ratio, Freeport Minerals was able to arrange the debt financing in a manner that minimized risk to the equity investors. By the time Ertsberg came on stream in December 1972, $146 million (including accrued interest) had been invested, of which $95 million was supplied by US insurance companies (with an OPIC guarantee), US commercial banks (with an Export-Import Bank guarantee) and KFW (in Deutschemarks); $24 million was provided by Japanese copper consumers and trading companies; and $24 million by equity. The trust agreement provided that all sales proceeds were to be paid to the trustee who would retain a portion equivalent to the principal and interest payments due at the next payment date, and the balance would be remitted to FI. The maturity of the loans was tied to the "completion date," related to a specific rate of production which was achieved at the end of 1973. The interest rate on bank loans was 0.5 per cent above prime rate, while other loans had fixed rates ranging from 7 to 9.25 per cent. The KFW loan bore the lowest rate of interest, 7 per cent. However, this turned out to be an expensive arrangement due to the subsequent appreciation of the DM in relation to the dollar during the period of the loan. Unlike most nonrecourse loans, Freeport Minerals was not liable for the entire indebtedness in the event the mine was not completed. Freeport Minerals was obligated to provide an amount equal to 20 per cent of its initial investment of $16.9 million in the event of cost overruns and also agreed to purchase a maximum of $9.1 million of the commercial bank loans if project completion was delayed.

As a result of various start-up and technical problems the mine did not operate satisfactorily until the end of 1974. Additional capital expenditures were necessary, but they were financed from FI's cash flow.

Since it was expected that the original Ertsberg open-pit mine would be exhausted by 1984, FI investigated a copper ore deposit about two-thirds of a mile east (1 km) of the operation, known as Ertsberg East. This deposit had reserves several times that of the original mine and was developed at a cost of over $100 million. Financing for Ertsberg East was provided by commercial banks and an insurance company, again without liability on the part of Freeport Minerals. It may also be noted that a substantial portion of Freeport Minerals equity investment in FI was guaranteed under OPIC's political risk insurance program, as were some of the commercial bank and insurance company loans.

The 1967 Contract of Work with the Government
The rather easy negotiation of the mining agreement or "contract of work" between FI and the Indonesian government reflected the new government's desire to attract foreign investment. Among the favorable provisions in the 1967 contract were:

(1) FI was given full control of exploration, design, construction, plant operation, marketing and the negotiation of long-term contracts. FI was only obligated to consult with the Ministry of Mines with respect to marketing the products and the prices paid under contracts "to the end that representatives of said Ministry shall be familiar with FI's marketing policies and procedures, and, if requested by FI, be able to assist FI in dealing with such marketing problems as may arise." Such language is indeed rare in mining contracts with governments.

(2) Except for resettling and compensating indigenous inhabitants in the project area, the use of the area was not subject to any rents and there were no royalties on output or export taxes and no import duties on items required for construction and operation of the project.

(3) FI was granted exemption from income tax liabilities for three years following the beginning of the 30-year operating period. Thereafter FI was to pay taxes on net income for the next seven years at 35 per cent, and after that at 41.75 per cent. However, during the first seven years of income tax liability, taxes paid were to be no less than 5 per cent of net sales during the tax year and thereafter no less than 10 per cent of net sales during the tax year.

(4) Taxable income was subject to the usual deductions of production costs, overhead, interest on borrowed capital, and depreciation and amortization of preproduction costs at a rate of 12.5 per cent per year.

(5) FI was given the right to retain the proceeds of sales in foreign currencies abroad and to use these funds for payment of debt service dividends, imports of materials, and other foreign claims.

(6) FI agreed to conduct a comprehensive training program for Indonesian nationals to be reviewed by the Ministry of Mines and to carry out an employment program so that "at least 75 per cent of all

positions in each employment classification are held by Indonesian nationals within eight years after the commencement of the operating period." However, FI's obligation to achieve these percentages was to depend upon the availability of Indonesian nationals with qualifications acceptable to FI.

As frequently happens following the construction of a project, particularly if initial earnings prove favorable, in 1974 the Indonesian government requested a renegotiation of the contract. There were several reasons for this. First, the government began negotiating what came to be known as Generation II agreements that were less generous than the Ertsberg contract of work. The Generation II agreements provided for land rents and royalties, the elimination of the tax holiday, higher income tax rates, and a reduction from eight years to five years in the period within which 75 per cent of all positions must be filled by Indonesian nationals. In addition, it required that 5 per cent of the existing shares in the operating company be offered to the Indonesian government in the sixth and tenth operating years. A second factor playing a role in the demand for revision of FI's agreement was the sharp increase in copper prices in 1973 and early 1974, which had the effect of increasing FI's net income to a relatively high level in 1974.

The government's request for modification of the 1967 agreement included provisions to bring the agreement more in line with Generation II contracts. The negotiations were friendly and compromises were reached that did not give the government all it requested. The tax holiday was reduced from three years to one year and the company had to pay a corporate income tax (beginning in July 1974) of 30 per cent for the first two years, 35 per cent for the next seven years, and 41 per cent thereafter. FI also arranged to sell the government, at book value, 8.5 per cent of the total equity shares in FI. However, no agreement was reached on several other government requests, including the provision of a royalty of 3.57 per cent on the value of contained copper in the concentrates. Although there have been problems with the government administration, such as obtaining work permits for foreign personnel, relations have generally been good. The president and several other top officers of the company are Indonesian, which has facilitated relations with the government. By 1981 100 per cent of the skilled, unskilled and clerical personnel were Indonesian; 90 per cent of the professional personnel and 75 per cent of the managers were Indonesian.

Financial Performance

Between 1973 and 1984, FI had a positive after-tax net income in every year except 1977. However, most of this income was used for reinvestment in the mine, so that by the end of 1984 retained earnings totaled about $101 million. Dividend payments were made in 1975 and 1976 and in each of the five years 1980–4. Over the entire period of operations through 1984, dividends totaled $83.5 million on an equity investment of $24 million. This constitutes an internal rate of return of 13.3 per cent. At the end of 1984 net working capital amounted to $56.5 million, and long-term debt had been reduced to $28.2 million (Freeport-McMoRan, *Annual Reports*). If the dividends paid by FI to its stockholders were to continue for another decade at the average level of the 1980–4 period, the IRR would be 17.3 per cent.

Conclusions
The decision of Freeport Sulfur (now Freeport-McMoRan) to make a large high-risk investment in Indonesia in 1970 was based on the following factors:

(1) the discovery of an orebody containing 2.5 per cent copper with gold and silver values, compared to an average grade of less than 0.7 per cent in the USA.

(2) the desire of the company to diversify its investments from heavy reliance on sulfur extraction

(3) a feasibility study showing a DCF of 40 per cent on equity on the assumption of 40 cents per pound of copper, a price that appeared conservative in view of the 1969 price of 47 cents per pound (US producer price) and a favorable market outlook

(4) a favorable investment contract with the Indonesian government

(5) a highly leveraged financial program with limited equity-investor risk beyond the initial investment.

Although the financial returns did not live up to expectations as a consequence of capital cost overruns, inflation and low real copper prices in the late 1970s and early 1980s, the project has proved profitable. Had the technical problems proved insoluble, or if initial earnings had not been higher than expected, or if the government had seriously violated the contract, the financial outcome would have been far less favorable or even disastrous.

Rosario Dominicana: Financial Success Invites Contract Renegotiation (Prepared by James E. Zinser, *Professor of Economics, Oberlin College*)

Resource companies take substantial risks when making investments in developing countries, and if they experience high profits, as a consequence either of increased product prices, or the discovery of larger than expected reserves, or for other reasons, the host government almost invariably demands a renegotiation of the contract calling for a larger share of revenues, or, in some cases, nationalization of the investment. Contract renegotiation or nationalization occurred in nearly all successful foreign petroleum investments in developing countries following the sharp rise in petroleum prices in 1973–4, and, as noted above, a demand for renegotiation occurred in the case of the Ertsberg mine. World prices of gold and silver rose almost as dramatically as the price of petroleum between the late 1960s and 1980, and Rosario Dominicana's contract experience with the Dominican Republic government parallels that of a number of petroleum companies during the 1970s.

The Pueblo Viejo mine developed by Rosario Dominicana SA during 1969–75 has a long history. Columbus found the Arawak Indians mining the area in 1494 and the deposit was mined intermittently by the Spaniards during the early part of the sixteenth century. Because of its low ore grade there was no further work on the mine until the 1950s when exploration and metallurgical testing were undertaken with unsatisfactory results. In December 1968 exploration rights were secured by Minerales Industriales and the following year control of exploration was acquired by the US mining firm, Rosario Resources (now a division of AMAX), which then formed a joint venture with Simplot Industries to explore and later develop the mine.

The early drilling results initiated in September 1969 were not encouraging, but further exploration led to the discovery in 1971 of an oxidized ore zone with indicated reserves of 6.5 million mt of oxidized gold/silver ore averaging 0.192 ounces of gold and 1.48 ounces of silver per ton. In addition to the oxide reserve, an estimated 17.7 million mt sulfide orebody averaging 0.13 ounces of gold, 1.12 ounces of silver and other values was discovered. Further drilling in 1972 revealed larger oxide reserves totaling 30 million mt averaging 0.15 ounces of gold and 0.76 ounces of silver. A mine and processing plant was designed by Fluor Utah Inc. with a capacity of 8,000 tons per day, capable of producing 350,000 ounces of gold and 1.5 million ounces of silver annually. It is reportedly the largest gold mine in the Western Hemisphere (Petrick Associates, 1980).

The initial financial plan for Pueblo Viejo provided for an equity investment of $1 million by Rosario Resources and Simplot Industries, with the remaining $20 million in development costs to be financed by loans from the Export-Import Bank and private US banks with an OPIC guarantee. This financing program did not materialize and in June 1973 the Dominican subsidiary formed to operate the project (Rosario Dominicana) negotiated an agreement with the Banco

Central of the Dominican Republic (as agent for the government) for a $30 million line of credit with local private banks to meet the revised $46 million investment cost (which increased to $47 million by the end of 1975). A 20 per cent interest in Rosario Dominicana was sold at cost to the government, with the remaining 80 per cent held by Rosario Resources and Simplot. The total equity was about $12 million.

In 1972 a mining concession was granted by the government under the 1971 mining code. There was a 42 per cent corporate income tax and an 18 per cent dividend remittance tax on net dividends. Assuming a 100 per cent remittance rate, the government's revenue would be 52.4 per cent of net profits. Earnings from the government's 20 per cent ownership of the subsidiary would increase its share of revenues to 62 per cent.

The mine became operational in April 1975 and in that year exported 270,000 ounces of gold and silver (in metal amalgam bars) worth about $28 million and yielding net earnings of $8.1 million for 1975 and $18.0 million for 1976. The free market price of gold was only about $36 per ounce in 1970 when the first investment was made, but by 1975 the average price of gold was $161 per ounce, eventually rising to over $600 per ounce in 1980. Meanwhile, Rosario Resources and Simplot had been exploring the Los Cacaos deposit adjacent to Pueblo Viejo and in 1975 petitioned the government for an exploitation concession on that deposit. The partners wanted the organizational structure of Rosario Dominicana to remain the same, but Pueblo Viejo's metallurgical plant capacity was to be increased to handle the ore produced by Los Cacaos. This petition aroused substantial criticism from government officials and subsequent negotiations eventually led to the loss of both Pueblo Viejo and Los Cacaos.

In response to the petition to the government for an exploitation permit for Los Cacaos, President Balaguer of the Dominican Republic appointed a "Gold Commission" to review the proposal. The commission's majority recommended either 100 per cent government ownership of Los Cacaos and a separate processing facility, or a revised Pueblo Viejo-Los Cacaos contract that included greater government ownership. Neither approach was acceptable to the owners of Rosario Dominicana, and in May 1976 Rosario and the Banco Central negotiated a preliminary agreement calling for the organization of a new company to develop Los Cacaos, in which the government would own 50 per cent, but Rosario and Simplot would provide the capital and would also increase Pueblo Viejo's plant capacity from 8,000 to 12,000 mt per day without otherwise affecting the structure of Pueblo Viejo. However, President Balaguer refused to approve the agreement and instead insisted that the Pueblo Viejo concession be revised to permit larger government participation and that Los Cacaos be developed solely by the government.

In response to a government demand for a renegotiation of the Pueblo Viejo concession agreement, an agreement was reached in December 1976 whereby the government's equity was increased from 20 per cent to 46 per cent, with the shares sold to the government at book value for approximately $7.4 million. The agreement also provided for an additional 5 per cent tax on net profits, the proceeds of which would be designated for local projects. The additional tax, together with the increased government participation raised the government's share in the earnings of Pueblo Viejo to almost 76 per cent.

During 1978 and 1979 rising gold and silver prices substantially increased Rosario Dominicana's after-tax earnings to $27.5 million and $47.9 million respectively, with a consequent increase in government pressure for a larger share of the profits. In 1979 the government imposed a tax on gold sales with the rate depending upon the world price of gold. In addition, it imposed a special net profits tax also based on the price of gold. Jointly these taxes had the effect of raising the government's share of net profits to almost 90 per cent for gold sold at prices above $300 an ounce.

The final action taken in October 1979 was an agreement by Rosario Resources and Simplot Industries to sell their remaining 54 per cent interest in Rosario Dominicana for $70 million, of which $24.1 million represented retained earnings reinvested in the corporation. For Rosario and Simplot the payment represented a long-term capital gain of $32 million, a return of 240 per cent on their equity. While this return appears generous, it was not unreasonable in view of the risk incurred by the investors in initiating the project. Moreover, the settlement was small compared to the expected value of future production. The foreign investors also lost their concession on Los Cacaos and their exploration outlays on that project.

Notes

1 Portions of the material in this chapter were taken from Mikesell, 1983, Chapter IV. Recent data and information were taken from Southern Peru Copper Company, *Annual Reports*.
2 This study is based in part on a case history in Mikesell, 1983, Chapter III.
3 For a history of the discovery and early exploration of Ertsberg see Wilson (1981). Wilson conducted the initial investigation of the orebody in 1960 and later was manager during the exploration period. The Ertsberg copper orebody was a dome-shaped mass rising over 500 feet above a narrow valley 12,000 feet high.

6

Government Policies and Regulations

Mining is subject to more government regulation than most industries. Since minerals tend to be found in remote mountainous areas, mines are frequently on government property. Public property has uses such as recreation that conflict with mining, which requires structures and activities that disfigure the terrain and pollute air and water. The latter create problems well beyond the mining area so that mining, smelting and other forms of mineral processing are subject to a variety of environmental regulations. Underground mining in particular involves substantial danger to workers so there are governmental regulations in the interest of mine safety.

Because minerals are basic raw materials in production, industrial countries have adopted programs for assuring their availability for domestic use. These programs may be designed to encourage domestic exploration and development, and to promote domestic self-sufficiency by limiting mineral imports, or by stockpiling for use during national emergencies. Conversely, governments of developed countries have sometimes discouraged investment in mining. The US government brought pressure on copper producers to reduce producer prices in 1965, and in 1972–4 domestic prices of copper, steel and other metals were kept lower than world prices by means of governmental controls. Price controls, subsidies and other interferences with the markets for metals tend to be destabilizing to the industry and impair the ability of US metal producers to modernize and become competitive with foreign producers. Some countries have discouraged investment in their mining industries by limiting or otherwise controlling foreign investment.

Minerals have traditionally been subject to international trade controls. As is the case with other primary commodities, mineral prices in world markets tend to fluctuate widely, since international demand and supply elasticities are quite low. Governments often seek to protect their domestic mining industries from losses due to world prices. This may be accomplished by import controls to maintain domestic prices and subsidies to make exports profitable in the face of low world prices. On the other hand, in periods of high world prices, governments often limit export of minerals to hold down domestic prices.

Adequate supplies of minerals in time of war are regarded by some governments as an important aspect of national security. In the nineteenth century access to minerals was often a reason advanced for acquiring colonies, or for preventing unfriendly powers from acquiring them. Access to minerals has continued to play a role in international conflict in the twentieth century, but is more important in petroleum than nonfuel minerals. Reduced dependency on foreign supplies of minerals is an important objective of national policy in some industrial countries, particularly the USA and the USSR.

It will not be possible to deal comprehensively with each of the above areas of government policies and regulations. Moreover, governmental objectives differ widely among countries, depending upon their mineral resource endowments and their industrial requirements. Therefore, only a few examples of each category of government policies will be given in this chapter. Since government controls on foreign investment in Third World countries were dealt with in Chapters 4 and 5, this chapter is mainly confined to mining policies in developed countries.

US Mineral Policies

Throughout the period following the Second World War, US mineral policies have been heavily oriented toward assuring the availability of minerals in time of war. Prior to the Second World War, the USA was self-sufficient in most major minerals. However, since the war the USA has become partially dependent on foreign supplies for basic minerals and increasingly dependent on minerals, such as bauxite-alumina, chromium and cobalt, that have never been produced in significant quantities in the USA, but whose use in industry has grown very rapidly. There have been two approaches to assuring availability of minerals in time of national emergency. One has been the establishment of a large national defense stockpile which, with few exceptions, is available for use only in a period of all-out war. The other approach has been to promote domestic production of minerals, including those that could not be economically extracted at current world prices.

The basic elements of US minerals policy are given in President Reagan's *Report to Congress on a National Materials and Minerals*

Program of April 5, 1982 – a report requested by Congress in the National Materials and Minerals Policy, Research and Development Act of 1980. This report recognized the role of minerals for both national production and national defense, but gave particular emphasis to the latter. As a means of expanding domestic output of minerals, the President's report gave considerable weight to increasing the accessibility of public lands for mineral exploration and development. While this expression greatly pleased the mining community, it disturbed environmental groups who oppose mining in wilderness areas, national parks and other public lands on which exploration and mining are currently restricted.

In his report, the President stated that his administration was considering the feasibility of using Title III of the Defense Production Act of 1950 (as amended) for subsidizing the creation or expansion of capacity of minerals that cannot be profitably produced at current world prices. Although there has been considerable congressional support for using the Defense Production Act for this purpose, the Office of Management and Budget (OMB) thus far has opposed government purchases of domestic materials at subsidy prices. OMB representatives have stated that the USA should rely on the market to improve the competitiveness of the American mining industry and that US national security objectives should be met by purchasing materials for the national strategic stockpile at world prices. Other US government agencies have favored a program of domestic subsidies.

Another element of US mineral policy mentioned in the President's report was the encouragement of materials research with a view to promoting technological innovations that would increase productivity and conserve the use of strategic minerals for which the USA is heavily dependent on foreign sources. However, the report failed to propose a specific program for governmental financial assistance for this purpose.

The major emphasis in the President's report was given to the strategic and critical materials stockpile (national defense stockpile). The present stockpile program dates from the passage of the Strategic and Critical Materials Stock Piling Act of 1946. This Act has been the subject of several amendments dealing with acquisition of strategic minerals, disposal of materials in excess of goals, and conditions under which stockpiled materials may be drawn down. The methods for determining stockpile goals for dealing with a wartime emergency have also been changed from time to time, together with the nature of the wartime emergency which the stockpile is designed to serve. In the past some administrations have been less favorable to the accumulation of a large strategic stockpile than others. The bulk of the stockpile accumulations occurred during the Eisenhower administration, but President Nixon favored reducing stockpile inventories by 90 per cent. In his 1982 report, President Reagan endorsed the national defense stockpile program formulated by Presidents Ford and Carter, which provided for a stockpile to meet military, industrial and essential civilian needs for the first three years of a war. (The aggregate value of the goals was about $17 billion in March 1984, of which only $11 billion was actually held in the stockpile inventory.) President Reagan ordered a reexamination of the stockpile goals and in 1985 approved a National Security Council report recommending a reduction in the aggregate value of the stockpile goal to $6.7 billion and no (net) additional stockpile acquisitions (White House, 1985; Mikesell, 1986a, p. 68). This appeared to critics as a repudiation of the President's policy outlined in his 1982 report.

The President's 1982 report was widely criticized as consisting of platitudes rather than coming to grips with conflicts, such as those involving land use, and of failing to propose specific programs, including recommendations for funding research and development and for meeting the national defense stockpile goals. The report tried to provide something for all groups with divergent interests in mineral programs, but failed to satisfy any group. This is not surprising. It is very difficult to formulate a satisfactory minerals policy and a program for its implementation because mineral issues encompass a wide range of commodities, each with special economic, military and social considerations and remedial options. Minerals such as iron ore, bauxite, cobalt, titanium and copper, to name a few, have too little in common in terms of their relationship to national concerns and objectves to be encompassed in a general statement on national minerals policy. Moreover, the major national policy issues are not oriented to minerals, but are concerned with the growth of the national product; holding down real prices of goods and services; protecting the civilian economy and national defense capability in periods of national emergency; and preserving the environment. Mineral policies and programs play a role in these major national policy issues, but are instruments for achieving goals rather than goals in themselves. US mineral policy is determined not by presidential rhetoric, but by what the government does regarding a large number of issues that concern minerals, such as taxes on mining, environmental regulations, the use of public lands, restrictions on foreign trade, and direct government assistance to mining. Existing programs tend to represent a compromise between contending interests rather than implementing a consistent minerals policy.

Most ores and concentrates enter the USA either free of duty or subject to low tariffs. Refined metals and alloys pay higher duties, but in general duties have not been high enough to prevent a decline in the domestic market shares of US firms producing copper, zinc, ferromanganese and other alloys. In recent years the US administration has rejected the demands of the mining industry for import protection for copper, zinc and other metals.

The USA encourages nonfuel mineral mining by means of tax incentives, the most important of which is percentage depletion. The depletion allowance (which is deducted for determining taxable income) is calculated on gross income at various percentages ranging from 15 per cent for copper, iron ore, gold and silver, to 22 per cent for lead, zinc, nickel and tin. The allowance may not exceed 50 per cent of net income, calculated without regard to depletion. Intangible costs of mining development can be expensed as well as exploration expenditures, up to a certain limit. The producer can elect full expensing of exploration, but must surrender his depletion allowance until taxes are equalized for the two systems. No special income tax rates are available for mining and there are no special investment credit allowances applicable to mining that are not available to other types of investment. In the past, the USA has provided some subsidies for certain minerals through Defense Production Act purchases, but not in recent years.

Vulnerability to Import Disruption of Strategic Minerals

The political unrest in central and southern Africa has heightened US concern for a possible disruption of supplies of strategic materials. South Africa is a major source of US imports of chromium, manganese and platinum, while Zaire and Zambia supply the bulk of its cobalt imports. The US national defense stockpile is not designed to protect the civilian economy against import disruption in peacetime. Releases from the stockpile are limited to military uses except during all-out war. Therefore, many believe that an economic stockpile should be created to deal with import disruption in peacetime, but such a program has never been favored by the US government. There is strong support in Congress for subsidizing production of low-grade domestic reserves of strategic materials, such as cobalt, but adequate stockpiles of imported minerals would be cheaper and more readily available in times of emergency. Stockpiling for domestic use would be economically feasible for only a few commodities, mainly chromium, cobalt, manganese and platinum. Other minerals on which the US is heavily dependent on imports are available from a number of countries not especially vulnerable to supply disruption (Mikesell, 1986a, Chapters 4 and 5).

Canadian Nonfuel Mineral Policies

Most of Canada's nonfuel mineral production[1] is for export, and in recent years mineral exports have constituted about 30 per cent of its total exports. Canadian policies are designed to promote exploration, development and processing of minerals by favorable tax treatment at the national level, by various types of financial assistance and by tariffs on imports of processed minerals. Much of the Canadian mining industry is foreign-owned. While such investment is welcome as a means of expanding the country's mining industry, Canadian policy is also directed toward encouraging majority Canadian mine ownership and control. Tax advantages are offered to both Canadians and foreigners who invest in Canadian-controlled corporations. In 1973 the Canadian government established a Foreign Investment Review Agency to review applications for foreign acquisition of Canadian business, new foreign investments, and expansion of existing foreign-controlled firms in new areas of business. In the nonfuel minerals area this regulatory program has not seriously restricted foreign investment. Canada has been more aggressive in encouraging increased Canadian ownership and control of the mineral fuels industry.

Canadian tax policies are complicated by the existence of both federal and provincial taxes. The combined tax policies have discouraged investment in mineral exploration and development in some of the provinces. The federal and provincial governments provide direct assistance to the mining industry through financial subsidies, exploration support, and encouragement of R & D.

As is the case in all industrial countries, Canada has clean air and water legislation that applies to mining and smelting. The Federal Territorial Lands Act of 1970 provides for conservation of lands in northwestern Yukon territories, including withdrawal of land from commercial use. However, the establishment of wilderness areas for the preservation of environmental amenities and recreational activities is not as important in Canada as it is in the USA. Canada has far larger land areas with low population density and little commercial development.

Nonfuel Mineral Policies of Australia

Australia's mining industry is similar in many ways to that of Canada. The bulk of the minerals produced are exported and these exports constitute a high proportion of Australia's export income. Also, a substantial portion of its mining industry is controlled by foreign capital. In general, Australia's policy has been to promote nonfuel mineral exploration and development, but the tax regimes and mining incentives have tended to change with the government in power, with the Liberal-National party being more favorable to the mining industry than the Labor party. Australia's foreign investment policy has been to achieve at least 50 per cent Australian equity and control of mining enterprises. Majority foreign-owned mining companies are not permitted to obtain contracts unless they provide a plan for a gradual increase in Australian equity to at least 50 per cent.

The Australian government has imposed controls on mineral exports in order to influence export prices and Australia participates in a number of international commodity groups, including the International Tin Association and the Bauxite Association. Australia also has sought to promote downstream processing by imposing high tariffs on processed minerals.

Australia requires that environmental impact studies be prepared before any new mineral ventures can proceed and requires environmental safeguards against air and water pollution. The government discourages mining in urban or suburban areas and no mines are permitted on lands reserved for the Aboriginals without the consent of their representatives.

South African Mining Policies

South Africa's investment tax and labor laws are designed to encourage mining activities. Most of the industry is domestically controlled, although in the early years of development substantial capital was obtained from the USA and Western Europe. South Africa requires a separate entity for the development and operation of each mine. This may have had the effect of discouraging majority foreign ownership in South African mining companies.

The government controls gold sales through a marketing sales company (INTERGOLD). All gold refining is done in a central facility which is effectively a cooperative which subsidizes high-cost producers in periods of low prices. DeBeers, a large mining and marketing group controlled by AAC, has a virtual monopoly on worldwide diamond sales and is a government-sanctioned monopoly.

South African taxes treat base and precious metals differently in that gold mines pay a supertax which may be as high as 65 per cent of net income. The tax regulations provide generous capital recovery and allow imputed interest on unrecovered capital as a cost that may be deducted in calculating taxable income. The South African tax code as applied to gold mining probably comes closer than any other developed country's code to taxing "economic rent."

Much of the gold mining is done in urban and suburban areas so that pollution is strictly controlled. Many mines are underground operations and employ thousands of people living in residential areas near the mines.

South African labor policy has provided a large pool of low-wage workers. The size of the pool has been increased by encouraging black workers from neighboring countries to work in South Africa on 6-month to 1-year contracts. Mines provide low- or no-cost housing and food for workers so that much of their earnings can be saved, and the government has not taxed these earnings. Normally, workers have sufficient cash at the end of their contracts to provide living expenses in their home country for one or more years. This arrangement makes it possible for workers to rotate between South Africa and their home country every other year.

Japanese Nonfuel Mineral Policies

Japan is heavily dependent upon foreign sources for nonfuel mineral ores and concentrates, but is a large producer and exporter of steel and nonferrous metals. Although Japan subsidizes its small domestic mining industry, its nonfuel minerals policy is oriented to the promotion of overseas exploration and development by Japanese firms or joint ventures with foreign firms, and to the negotiation of long-term contracts for foreign supplies, often combined with loan financing for their development. In implementing this policy there is close collaboration between Japanese government agencies and private firms. Associations of private firms, with financing provided by the Industrial Bank of Japan, stockpile copper, zinc and aluminum. However, the stockpile program is not oriented to a future war, but to stabilizing prices to domestic users.

Japanese tariffs on ores, concentrates and other unrefined nonfuel minerals are minimal or zero; tariffs on refined metals, including aluminum, copper, zinc, lead, nickel and titanium are high relative to those of other countries. Thus, the Japanese tariff system is designed to protect domestic value-added in nonfuel mineral processing.

Nonfuel Minerals Policies of the EEC Countries

The EEC countries are less dependent on imports of nonfuel minerals than Japan, but much more dependent than the USA. The common EEC tariff provides for low or zero duties on imports of ores and concentrates, but relatively high duties on refined and processed minerals. The Lomé Conventions between the EEC and sixty-one developing countries that were previously dependent territories of EEC countries, provide for various types of assistance to the mining industries of these countries, and tariff preferences and export stabilization arrangements.

France
France has the most active and comprehensive government program for promoting domestic overseas nonfuel mineral supplies. The Bureau des Recherches Géologiques et Minières (BRGM) is the principal vehicle for implementing the government's nonfuel mineral policies. It promotes exploration and mining in some forty countries, both directly through its own subsidiaries and in collaboration with private companies, and provides technical and financial assistance to foreign SMEs. The French government also provides tax incentives and financial assistance for development of domestic minerals. The objective is to expand and diversify foreign and domestic sources of nonfuel minerals. The government has a modest stockpile which is limited both in terms of the number of commodities and of volume of inventory. France is a member of several international commodity stabilization associations.

Germany
Germany supports domestic and overseas mineral exploration and development, but relies mainly on private industry rather than government agencies for implementing this policy. The government promotes overseas mineral exploration by grants to German mining companies and the BGR (a government technical institution) conducts geological surveys directly and in collaboration with German mining companies or foreign governments. Overseas mineral development is promoted by investment guarantees and long-term credits. The DEG (German Association for Economic Cooperation) is a government-owned institution organized to acquire equity interest in, and provide loans to, raw material ventures in developing countries.

The German government generally opposes international commodity agreements and other arrangements that interface with free markets for minerals. In the past it has also rejected the concept of a government-sponsored nonfuel minerals stockpile. The Ministry of Economics has provided grant assistance for domestic exploration projects, and the government also participates directly in R & D related to the nonfuel minerals industries.

United Kingdom
British government activities in both domestic mineral resources and the development of overseas supplies are quite limited compared to

those of Germany and France. The UK has no direct programs for the support of overseas exploration and development by private concerns. It does provide political risk insurance, trade credits and some tax advantages for UK mining corporations. There is also limited assistance to domestic exploration. Britain maintains no nonfuel minerals stockpile.

Nonfuel Minerals Policies of Developing Countries

The policies of mineral-exporting developing countries are oriented to maximizing foreign exchange earnings from their resources and making the mineral resource industry contribute to the economic, social and political goals of the government. These objectives are frequently in conflict. As noted in Chapter 4, this is especially true where governments discourage foreign investment in order to promote government or private domestic control over resource industries. The mineral industries of most developing countries are a mixture of SMEs and foreign investment, plus some private domestic ownership, usually in the form of small mining enterprises. Although SMEs are often provided with certain tax advantages and access to public lands and credits, they tend to operate as profit-maximizing entities similar to privately-owned firms. Private domestic mining enterprises frequently receive government assistance in the form of credits and tax advantages. Chile, Indonesia, Malaysia and Peru are good examples of countries with strong SMEs and domestically owned mining sectors. In Brazil, SMEs dominate the mining industry, but there is a strong private domestic sector; many foreign investors in the resource industries are in joint ventures with government or private enterprise. In Mexico and the Philippines, the domestic private sector dominates the mining industry, but there is considerable foreign minority participation. In Zaire and Zambia, large-scale mining is wholly controlled by SMEs. In PNG the mining industry is controlled by foreign private companies, but the government maintains a substantial degree of surveillance and has minority equity participation. In India there is relatively little foreign investment in nonfuel minerals, but there is a strong private domestic sector, especially in iron ore.

Taxation of Mining in Developed Countries

Governments impose a variety of levies on mining operations. There are essentially three forms of taxes: (1) those based on mine output; (2) those based on net income; and (3) those based on property or the use of state-owned land. The oldest form of taxation on mining is the royalty, which may be a certain amount of money per physical unit of output or a certain percentage of the value of mine output. Virtually all countries have a tax on net income and, in some cases, special tax rates or methods of calculating taxable income are applied to mining. Nearly all countries have property taxes on privately owned land and other physical assets, but there may be a special method of valuation or tax rate applied to mining property. Where the government owns the land or the minerals in the subsoil, it may lease the land to the mining company. In some cases a bonus may be paid to the government for the right to mine a particular area instead of, or in addition to, annual lease payments.

Developed countries tend to use the same basic tax structure for mining as is used for other industries. Since most mining firms are domestically owned, governments are reluctant to favor one industry over another unless there are important social or national security reasons for doing so. The objective of the government is not to maximize its revenue from any particular industry, but to administer a fair and equitable tax system without discriminating against any industry or region. The greatest differences in tax treatment between extractive and other industries may be in the treatment of depreciation. Mine deposits are depleting assets while buildings and machinery are depreciating assets; different depreciation principles are applied to the two types of assets. Also, developed countries do not usually discriminate in tax treatment between domestic and foreign investors, and there are tax treaties that proscribe such discrimination.

The US government does not impose royalties or lease payments on most nonfuel minerals extracted on public lands.[2] Under the Mining Laws of 1872 and 1976, miners can file claims on tracts of public land and acquire exclusive rights to explore for minerals and extract them without payment of any royalty or fee. Mining claims in wilderness areas are restricted, as are areas under study for possible wilderness designation. The principal source of federal income from mining is the corporate income tax and, as has been noted, mining is favored by the percentage depletion allowance. There are also state and local income and property taxes, and some state governments impose a severance tax (royalty) based on a percentage of the value of output.

In Canada, publicly owned lands are leased by the federal and provincial governments for mining, and there are also royalties. Both the federal and provincial governments levy income taxes. In Australia royalties are imposed by the states regardless of the ownership of the surface lands. Australia's federal government imposes a corporate income tax, but income from gold mining is not subject to this tax.

There exists a broad variety of tax systems applied to mining in Western Europe, depending in part on whether the minerals in the subsoil belong to the state (as is the case in France), or to the owner of the surface land (as is the case in Britain). In general, the major source of revenue to the government is the corporate income tax, but in some countries royalties are paid to the federal or provincial government.

International Trade Policies

International trade policies relating to minerals differ among countries with the structure of their mineral industries. In the case of countries that are large importers of a particular mineral, there has long been a tendency to maintain low or no tariffs on ores or concentrates, and relatively high tariffs on processed and fabricated minerals. Since the Second World War, developed countries have entered into a number of tariff-reduction agreements, and there exists a basically low tariff structure in industrial countries. The rules of the General Agreement on Tariffs and Trade (GATT) on the use of quotas and other nontariff barriers (ntbs) are widely violated, but nevertheless they impose constraints on the use of ntbs because of the likelihood of retaliation. Some countries, such as the USA, produce both mine products and processed and fabricated mineral commodities; this condition imposes additional constraints on the use of trade barriers.

During the 1980s the US copper mining industry has been depressed as a consequence of the low rate of growth in world demand for copper, world overcapacity in copper production, and rising imports from low-cost foreign producers. In 1984 US copper producers sought a quota on imports under the Trade Act of 1974, but the International Trade Commission (ITC) recommended against raising import barriers, partly on the grounds quotas would be ineffective in assisting the domestic copper industry. The reason is that by raising domestic copper prices above world prices there would be a large increase in imports of fabricated copper products and a decline in US exports of these products. This would have the effect not only of impairing the US copper fabricating industry, but that industry would reduce its demand for domestic copper, thereby offsetting any effect of a quota on copper imports (International Trade Commission, 1984).

This experience points to an important lesson for any country that is both a producer of a raw material and a fabricator of that material. A rise in the domestic price of a raw material above the world price will make its fabricating industry noncompetitive. The same principle applies to steel, aluminum and other metals that are used in manufacturing.

Where a domestic industry is highly integrated and produces some of its mineral raw materials in foreign countries, imports of raw materials are likely to be low or nonexistent. US steel firms obtain much of their iron ore from captive foreign mines; therefore, the steel industry is opposed to import restrictions on iron ore, but strongly favors barriers to foreign steel. Likewise, the US aluminum industry obtains most of its bauxite and alumina from foreign sources in which US aluminum companies hold equity interests. Thus, both iron ore and alumina are imported duty free. The US tariff on aluminum metal is only 2.6 per cent and the industry has little interest in raising the duty. This may be true in part because foreign production of aluminum by US firms supplies their fabricating facilities in the USA, and in part because a domestic price for aluminum significantly higher than the world price would impair the world competitive position of US fabricated aluminum products.

A number of developed countries, including the USA and the EEC, have granted tariff preferences to imports from developing countries, and, with certain exceptions and limitations, imports of minerals from developing countries are free of tariff duties, even in cases where they compete with domestic production. Tariff preferences also apply to processed and fabricated metals from developing countries. On the other hand, developing countries tend to maintain high import duties and other forms of restrictions on imports of processed and fabricated metals for which they have developed facilities to serve the domestic market. Developing countries are generally not constrained by the GATT code of fair trade practices.

Environmental Regulations

Mining is a dirty industry and creates a number of environmental problems. Mining, especially open-pit mining, deforms the surface of the land and creates large piles of unsightly waste materials which may contain hazardous substances that pollute water and soil. Water from mining and concentrating operations may contaminate the subsoil and the rivers into which it flows. Most serious of all are the gases produced by smelting, which may not only contaminate the air in the region of the smelter, but affect lakes and vegetation hundreds of miles away through the creation of acid rain. Since mines are frequently located in mountainous areas, their development impairs the environmental amenities enjoyed by recreationists. Also, due to the fact that exploration and mining frequently occur on publicly owned lands, there may be conflict between mining companies and environmentalists over the use of those lands. For these reasons it is not surprising that in nearly all industrial countries, and in some developing countries, mining and smelting are subject to a substantial amount of environmental regulation.

Virtually all industrial countries have environmental regulations and legislation for maintaining clean air and water standards. However, the environmental problems are so broad and complex that no country has dealt with all the problems in a manner satisfactory to those sectors of the public especially concerned with them. Governments are subject to pressure from the mining industry to take into account costs of various forms of pollution abatement. In the USA, legislation applies different standards to existing smelters (or additions to capacity of old smelters) and to new smelters. In some cases the requirements are in terms of equipment employed rather than amounts of pollutants emitted. Sometimes firms are given a certain number of years before specified equipment must be installed, or certain pollution abatement standards must be met. Standards differ considerably among countries and there are also differences in the way standards are administered. These differences affect the relative costs of pollution abatement among mining firms in different countries. In addition, there are differences among countries in the subsidy and tax programs employed to promote the installation of equipment designed to reduce pollution.[3]

The principal impact of pollution abatement regulations on the mining industry arises from regulations on emissions of SO_2 and other air pollutants from copper, lead and zinc smelters. Regulations include both ambient air quality standards and discharge limits for individual sources. Ambient air quality standards will differ with the environment in which smelters are located so that emissions in areas of high population concentrations must be less than those in sparsely populated areas. In the USA, discharge limits for new facilities, including additions to new facilities, are based on "best available control technology" for the permanent reduction of emissions. In other countries, firms are often permitted to use supplementary control strategies, such as high smokestacks and intermittent production, which involve lower costs than permanent control strategies requiring extensive modernization and new smelting technologies. When EPA regulations were first introduced in the 1970s, most US copper smelters were old reverberatory types that were required to be replaced within a certain period of time. In Japan and Europe smelters tended to be newer and less polluting. Also in Japan and Western Europe, it is possible to capture SO_2 smelter emissions in the form of sulfuric acid which can be sold at a price that covers cost of capture. In the USA, which is endowed with natural sources of sulfur as well as supplies from natural gas and crude oil desulfurization, the market for sulfuric acid produced by smelters is quite limited. For these and other reasons US smelters have been unable to dispose profitably of smelter-produced sulfuric acid output.

A comparison of costs of environmental regulations per unit of mineral products among countries is exceedingly difficult, especially since they may differ considerably within a country depending upon location of mines and smelters. Air quality standards are more strict in Japan than in any other industrial country because smelters operate in areas of high population density. They are less strict in Canada where mining tends to be located in mountainous regions with low population density. On the other hand, Japan appears to provide somewhat greater assistance in helping its smelting industry meet air pollution standards than does the USA.[4] It does appear, however, that EPA regulations have been an important factor in the reduction of US smelter capacity, in part because of the existence of old equipment. If US copper firms had been more profitable, they could have afforded to replace old capacity with new smelters that would meet EPA standards. It is reckoned that by 1990 there will be only four copper smelters in the US with estimated capacity of 575,000 st. (This excludes the hydrometallurgical facilities that are becoming increasingly important in the US copper industry.) US smelter production of copper in 1981 was about 1.5 million st (Siedenburg, 1985, p. 21).

Access to Public Lands

Although mining impairs environmental amenities wherever it exists, the conflict between mining interests and environmentalists is greater in the USA than in any other country. The principal reason is to be found in the high proportion of federally owned lands in areas where minerals tend to be found, and the strong public interest in wilderness preservation that has been growing throughout the twentieth century. As of 1979, the US government controlled about 760 million acres (about half of it in Alaska), or about one-third of the land in the USA. Approximately 42 per cent of these acres has been completely withdrawn from mineral activity, another 16 per cent is severely restricted and 10 per cent is moderately restricted (Comptroller General, 1979, p. 19). The major areas at issue between environmentalists and mining interests are some 80 million acres designated as wilderness under the Wilderness Act of 1964, plus 174 million acres identified by the Bureau of Land Management (BLM) as wilderness study areas (WSA) that are being reviewed for possible inclusion as designated wilderness areas. (Exploration and mining activities are not allowed or are severely restricted in wilderness areas and national parks.) The US mining industry favors elimination of such restrictions as well as limiting the acreage of new areas brought into the wilderness system. Environmental groups, such as the Sierra Club and Wilderness Society, seek substantial additions to designated wilderness areas and the elimination of all mining activities in or near the boundaries of wilderness areas and national parks.

The arguments of the mining community for maximum access to public lands for exploration and mining are couched in terms of the importance for national security of a strong and growing minerals industry and the contribution of mining to employment and national income. Environmentalists, on the other hand, argue that the limited amount of land with wilderness characteristics in the face of a growing demand for wilderness amenities renders the social value of these amenities higher than the commercial value of their mineral producing potential. A rational solution for determining alternative uses of public lands could be achieved by the employment of social benefit-cost accounting, but rational approaches usually give way to political compromises and there are powerful political forces mobilized on both sides (Mikesell, 1987).

Environmental Controls in Third World Countries

Environmental standards in Third World countries are generally much lower than those in developed countries. Indeed, governments of some developing countries have ignored environmental impacts almost entirely. The obligations of multinational mining companies in developing countries are generally set forth in the mining agreement. A review of such agreements over the past decade revealed only one agreement containing more than a perfunctory clause with respect to environmental protection, namely, the agreement for the Ok Tedi gold/copper mine in PNG. Although some multinational mining companies have employed environmental programs superior to those required by the host government, in general environmental standards of MNCs have been adapted to the particular public policies of the countries in which they operate.[5] Nevertheless, affiliates of multi-nationals use environmental standards that are superior to those employed by domestically owned firms in the same industry; this is especially true for SMEs and petroleum enterprises.

The willingness of a developing country to accept more air, water and land pollution than a developed country has sometimes been regarded as a factor in its comparative advantage in mining. For this reason it has been suggested that multinational mining firms are more willing to invest in a country with lax environmental standards. For example, a Chilean official reportedly stated to a prospective investor that Chile did not care how dirty smelters are that operate in desert areas of northern Chile. However, the alleged advantages of lax environmental standards have been challenged. First, there is little evidence that MNCs deliberately locate in so-called pollution havens. Currently more FDI flows into the USA than into any other country, and much of this investment is in pollution-intensive metals and petroleum. Yet US pollution abatement regulations are among the most stringent. Second, some multinational mining companies believe it is to their advantage to maintain environmental standards approaching those mandated in developed countries. Since mining companies tend to operate for a number of decades in a community, they want to maintain the respect and good will of the community. Moreover, Third World countries are in the process of raising their environmental standards and it is far less expensive to design projects that limit environmental impacts than to undertake extensive changes in project design a decade after construction. This applies particularly to air pollution, but is also important for mine waste.

The negotiation of the agreement between the Broken Hill Pty (BHP) mining consortium and the PNG government for development of the Ok Tedi mine provides a good illustration of the importance of undertaking environmental impact investigations as a part of the feasibility study for a mine. The PNG government was quite sensitive to the environmental impact of the project despite the fact that the mine is located in a remote area inhabited by Stone Age people in a few isolated villages. One reason was the environmental consequences of the Bougainville copper mine in PNG where mine tailings had been dumped directly into the Jaba River and had seriously disrupted both subsistent fishing and the natural ecology of the river. The PNG government held that its approval of the Ok Tedi project must encompass environmental aspects based on an adequate impact investigation. The BHP consortium – which includes Australian, German and US mineral firms – argued that it could not afford large outlays for environmental studies prior to the approval of the project. The limited environmental work done during the feasibility study proved inadequate for generating information needed to assess the impact of dumping waste material into the adjacent river system and of releasing cyanide effluent from the gold/cyanide plant. The PNG government refused to accept the conclusions of the BHP consortium that these activities would not cause environmental damage. There were extended negotiations and further studies by both the consortia and the government before a final compromise was reached (Pintz, 1984). These and other environmental issues contributed to the delay in final approval of the project.

The existence in the USA of environmental regulations on mining more stringent than those applied in other countries has been regarded as a source of competitive disadvantage for the US mining industry. Bills have been introduced in the Congress calling for special import duties that would equalize environmental costs. Discriminatory import barriers for this purpose would constitute a violation of the GATT and are generally rejected by economists (Walter, 1975). Environmental standards constitute part of the social policies of governments – similar to minimum wage, health and safety requirements. Import duties designed to equalize the cost impacts of differing social policies would subject world trade to a maze of discriminatory restrictions. Moreover, experience has shown that it is exceedingly difficult to make international comparisons of environmental costs.

Governments of most countries are becoming more conscious of the importance of preserving the environment and the gap between environmental standards of developed and developing countries will probably narrow. The World Bank and other international development institutions have mandated environmental impact studies for all projects they finance. Environmental requirements will inevitably be extended to domestic mining and metallurgy projects, including government projects. This will tend to equalize pollution abatement costs among mineral producing countries of the world.

Notes

1 For a discussion of Canadian mineral policies and those of other industrial countries discussed in this chapter, see Economic Consulting Services (September 1981).
2 There are oil and gas leases plus a few nonmetallic minerals such as phosphate and potash.
3 For a discussion of international environmental policy, see Walter (1975 and 1976).
4 For a discussion of differences in environmental regulations of major industrial countries see Congressional Budget Office (1985, Chapter 3).
5 For a description and analysis of environmental programs of MNCs see Gladwin (1985); and Gladwin and Welles (1976).

7

International Problems and Agreements

As is the case with trade, finance and nuclear materials, mining and minerals are the subject of many international problems that have been dealt with in international conferences and negotiations. A large portion of nonfuel minerals output enters into international trade and fluctuations in prices affect the welfare of many countries. Reserves of some minerals are highly concentrated in a few countries and there is concern that these countries may exploit their positions by forming cartels. Exploration has revealed a vast storehouse of minerals in seabed nodules and an effort is being made to establish international control over development and marketing of these resources. Finally, international development institutions, such as the World Bank, have been making loans to Third World countries for the development of their mineral resources during a period of overcapacity and low mineral prices. This assistance has given rise to strong objections by mining firms and by the US government.

International Commodity Agreements

The prices of most nonfuel minerals are subject to a high degree of short-term fluctuation. This arises because the demand for minerals tends to be highly inelastic, while productive capacity changes slowly in response to market prices. The degree of price fluctuation for an individual mineral depends in considerable measure on the nature of the world market. Where there are a number of producers that do not adjust output to changes in market prices, price fluctuations may be very large. For example, in the case of copper, lead and zinc the average annual percentage deviation of real prices from a 5-year moving average (in constant 1981 dollars) ranged from 14 to 15 per cent over the 1955–81 period. In the case of nickel and aluminum, the average annual percentage deviation over the same period was only

Table 7.1   *Average Annual Percentage Deviation of Real Prices from Moving Average, 1955–81*

Copper	15.3
Lead	14.7
Zinc	14.4
Tin	8.1
Manganese	7.2
Iron ore	5.7
Bauxite	5.6
Nickel	4.6
Aluminum	4.6

Source: World Bank, 1983, p. 45.

4.6 per cent (see Table 7.1). A high proportion of the world output of copper, lead and zinc is produced in developing countries and production in these countries is relatively insensitive to price fluctuations. On the other hand, a relatively high proportion of aluminum and nickel production is in developed countries and a few firms maintain substantial market control. The average annual percentage fluctuation of bauxite and iron ore prices was 5.6 and 5.7 per cent respectively over the 1955–81 period. In the case of bauxite, most production is in developing countries, but there are virtually no free markets, and price control is maintained by a handful of integrated aluminum companies. A similar situation exists for iron ore, much of the production of which is controlled by integrated steel companies.

Real mineral prices are also subject to long-run movements in the course of which prices are often well above or below the average full production costs. Unlike annual crops of wheat or cotton, the output of which usually adjusts from year to year with price changes, mine capacities cannot readily be adjusted to price movements except over the long run. Long-run price movements have been more serious for the mining industry than short-term price movements. In recent years there has been a bias toward overcapacity in the world mining industry. This has resulted in part from the declining rates of growth in world consumption of major metals as noted in Chapter 1. The relatively low level of investment in capacity during the past decade is likely to be followed by a period of high mineral prices as world consumption catches up with productive capacity.

For nearly four decades developing countries have campaigned for the creation of international commodity agreements designed to stabilize or maintain prices of primary commodities. At the end of the Second World War the USA and other industrial countries also favored international commodity agreements and the 1948 Havana Charter for the International Trade Organization set forth certain principles for the establishment and operation of such agreements. During the past two decades the developing countries, through the United Nations Conference on Trade and Development (UNCTAD), have been quite active in promoting commodity agreements while many industrial countries have been unenthusiastic about the feasibility and desirability of such agreements.

The only international commodity agreement for a nonfuel mineral is the International Tin Agreement (ITA), but there are associations of exporting countries for copper, bauxite and iron ore. The most active of these are the Intergovernmental Council of Copper Exporting Countries (CIPEC) and the International Bauxite Association (IBA). However, neither group is capable of exerting monopoly power in the production and marketing of its commodity.

UNCTAD Integrated Program for Commodities
The effort by Third World countries to establish international commodity agreements had its principal expression in the Integrated Program for Commodities (IPC), which was approved by both developing and developed countries at the fourth UNCTAD conference, in Nairobi in 1976. The IPC calls for the negotiation of buffer stock-type agreements for ten core commodities, of which copper and tin are the nonfuel minerals. For seven additional commodities, including bauxite and iron ore, other types of price stabilization measures would be employed. Membership in buffer stock agreements would include both producing and consuming countries, with the largest share of financing contributed by developed countries. In addition, the IPC provides for a common fund of $6 billion to be subscribed by governments to serve as a resource pool for facilitating the operation of individual agreements. At the time the IPC was formulated, international commodity agreements for four of the ten core commodities – cocoa, coffee, sugar and tin – were already in operation, and the USA has been a member of the coffee and tin agreements from time to time.[1] UNCTAD made several attempts to establish an international buffer stock agreement in copper, but was unsuccessful in obtaining financial and other support from the USA and certain other industrial countries.

The USA and other developed countries have in the past favored the concept of commodity price stabilization in the sense of reducing price fluctuations above and below a reference or target price which should approximate the long-run equilibrium price of the commodity. This objective can best be achieved by a buffer stock arrangement provided the buffer stock contains sufficient quantities of both money and the commodity to maintain the market price within a fixed margin above and below the reference price, and provided the reference price reflects the long-run equilibrium condition. Unlike export and import quotas, which are used in the international coffee agreement, a pure buffer stock arrangement does not interfere with the freedom of the market or encourage producers to accumulate inventories in excess of export quotas.

The UNCTAD Secretariat and Third World countries have favored the use of export quotas to supplement the operations of a buffer stock and have sought to maintain for the commodity a target price that producers regarded as "fair", or equal to some historical level, rather than using a reference price approximating the long-run equilibrium price. The developed countries, on the other hand, have insisted that any new commodity agreement avoid the use of quotas and other interferences with normal market operations, and that realistic reference or target prices be employed. Thus buffer stock operations would seek to reduce the amplitude of market price fluctuations without attempting to maintain prices at variance with the long-term equilibrium price. This difference in approach between developed and Third World countries has been a major obstacle to the negotiation of commodity agreements.

Commodity specialists and economists have disagreed regarding the feasibility of an international buffer stock operation.[2] First of all, it is extremely difficult to project a long-run equilibrium price that would serve as the reference or target price for stabilization operations. Long-run demand and cost conditions are subject to change for a variety of reasons that cannot be predicted. In practice, target prices have been negotiated by political representatives of producing and consuming members of the agreement, rather than on the basis of independent judgments by commodity specialists and economists. In the past, such politically determined targets have usually turned out to be too high, so that funds in the buffer stock were depleted in an effort to support a price above the long-term equilibrium level. In order to assure the success of the agreement, the buffer stock manager should be given authority to change the target price and the upper and lower price support levels from time to time, but granting such authority may be politically impossible. A successful buffer stock operation would be indicated by the reestablishment of the initial composition of the buffer stock in terms of money and commodities at some time over a period of, say, three to five years.

The ITA operated by the International Tin Council (ITC) is an example of the difficulties in achieving mineral price stability. Efforts at international control of the price of tin have had a long history.

The first ITA operated during the period 1956–61, while the sixth ITA came into force in 1982 but broke down in October 1985. Governments of twenty-two countries are members of the sixth ITA, of which six are major tin producers and sixteen are consumers. However, two large producer countries, Bolivia and Brazil, are not members, and the USA joined the fifth ITA, but not the sixth. The floor price at which the buffer stock manager buys tin and the ceiling price at which he sells tin were established by negotiation and the floor and ceiling prices have been changed from time to time. Both producers and consumers contribute money and tin to the buffer stock. In addition to buffer stock operations, a producer group – Malaysia, Indonesia, Thailand – employed export controls, and Malaysia has engaged in unilateral market support to boost tin prices.

Although the ITC has been able to maintain prices within the ceilings for substantial periods of time, prices above the ceiling for one to three months occurred seventy-seven times during the period 1971–9. However, until 1985 the market price of tin did not fall below the floor price. Between 1956 and 1979 the ITC changed the floor and ceiling prices eighteen times. The range between the floor and ceiling price during the 1970s varied between 20 and 26 per cent of the floor price. During the 1970s the floor price varied from £580 per long ton to £1,350 per long ton. Thus the overall stabilization record has not been especially favorable even with the aid of export controls.

Despite the low rate of growth in demand for tin, the ITC maintained prices ranging from $7.60 per pound in 1981 (annual average) to $5.55 per pound in 1984. This compared with $3.10 and $3.46 per pound in 1976 and 1977, respectively. Thus while prices of most other metals were declining in the mid-1970s and the early 1980s, tin prices rose significantly as a consequence of buffer stock purchases and export controls. The ITC borrowed over $420 million (secured with tin) to supplement the funds contributed by members to support the price of tin. By October 1985 the ITC had accumulated 65,000 mt of tin, which was 42 per cent of annual consumption of 155,000 mt in 1984. Production in that year was 175,000 mt (*Wall Street Journal*, 1985b, p. 6). In October 1985 the buffer stock ran out of cash and prices on the LME plunged. As a consequence, trading in tin on the LME was halted. In April 1986 tin sold at $2.70 per pound, rising to $3.45 in June 1986, well below the mid-1985 support price of $5.50. The sharp drop in prices was due to the threat that lenders to the ITC would sell the tin securing the loans.

Although the stated purpose of the ITC was to "stabilize" tin prices, in practice it sought to keep them as high as possible and raised the price floor in each new 5-year agreement. High prices encouraged consumers to use cheaper substitutes and encouraged other non-ITC countries to produce more tin. Brazil, for example, more than doubled its tin production from 1980 to 1984. The legacy of the ITA is overcapacity in the world industry and a long period of low prices before equilibrium between full production costs and prices can be restored.

Another obstacle to the creation of a successful buffer stock operation is the very large amounts of money and commodity that would be necessary to assure that the manager would not run out of either asset over a period of several decades of cyclical price movements. Simulation studies for the operation of a buffer stock agreement in copper have shown that a stock capable of achieving reasonable price stability is likely to require a combination of several billion dollars and several million tons of copper. Even if the required sum of money could be obtained from developed countries, there would be no way of accumulating the initial stock of copper without driving up the price to a very high level (Charles River Associates, 1977, pp. 1–6). Other technical studies on the feasibility of a copper buffer stock have reached more optimistic conclusions (Maizels, 1982). However, given the depressed copper prices during the first half of the 1980s together with uncertainties regarding the future rate of growth in demand for copper, agreement on a reference price around which to stabilize prices would be exceedingly difficult to reach. It is also unlikely that the US government would contribute substantial funds to an international commodity program, or that it would contribute to the $6 billion fund for the IPC common fund. At the present time there is little likelihood that an international copper agreement will be negotiated and the outlook for creation of an international price stabilization program in other nonfuel minerals appears even less likely.

Cartels for Nonfuel Minerals
Spokesmen for Third World mineral-producing countries have sometimes threatened to create producer cartels if they fail to establish price stabilization agreements involving both producing and consuming countries. Following the action of the Organization of Petroleum Exporting Countries (OPEC) in raising oil prices more than ten-fold during the 1970s, there was widespread fear that cartels might be formed in several nonfuel minerals and that they would be successful in raising prices of bauxite, copper and other metals. However, these fears have proved unfounded because the market conditions that existed in petroleum do not exist in nonfuel minerals. Petroleum was undoubtedly undervalued in 1972 since world demand was increasing at about 8 per cent a year while world petroleum reserves were not expanding rapidly and were actually declining in

the USA and other non-OPEC countries. A substantial rise in petroleum prices would have occurred by the end of the 1970s even in the absence of the OPEC cartel. Moreover, OPEC petroleum-producing capacity was very large relative to that of the rest of the world and OPEC members were successful in achieving a considerable degree of cohesion. Furthermore, one OPEC member, Saudi Arabia, was in a position to exercise a major influence on the world price by curtailing output. None of these conditions exist in any nonfuel mineral.

The rate of growth of world consumption in nearly all nonfuel minerals declined during the 1970s and by the second half of the decade there was considerable overcapacity in most metals. Moreover, in only a few commodities was over half the output produced by Third World countries, and in these cases there was not sufficient political cohesion to form a cartel capable of controlling the world price for a protracted period of time. The IBA was successful in raising export taxes on bauxite several-fold during the mid-1970s, but some members (including Australia) did not follow the lead of Jamaica in raising export taxes. Although the production of chromium, cobalt and manganese is heavily concentrated in a relatively few countries, political and market conditions do not favor the creation of cartels and, in addition, a sharp rise in the price of these commodities would soon promote the development and use of substitute materials.

Vulnerability to Foreign Supply Disruption

There are two types of foreign supply disruptions to which importing countries may be vulnerable: (1) those occurring during a global war when large portions of the world may fall under enemy control and world shipping from certain areas be cut off or greatly restricted; and (2) disruptions arising from civil disturbances, limited wars, or politically motivated embargoes affecting major supply sources of particular minerals. The first type of supply disruption occurred during the Second World War when supplies of materials such as rubber, tin and quinine were no longer available to the Western world because production was largely concentrated in Japanese-occupied territories.

The second type of disruption has not occurred on a significant scale in recent years, but there are widespread fears that disruption of supplies of chromium and manganese may occur due to civil disturbances in South Africa, or of cobalt due to political events in Zaire and Zambia. The 1978 invasion from Angola of Shaba Province in Zaire reduced cobalt production for a time, but actual shipments continued. Nevertheless, following the invasion, speculative activity resulted in a several-fold increase in the price of cobalt. The only case of a political embargo was that attempted by the Arab oil producers in 1973–4 when they embargoed oil shipments to the USA and the Netherlands. However, the embargo was not effective because once oil left the Middle East it was impossible to control its ultimate destination.

In the case of most nonfuel minerals, world supplies are sufficiently diversified so that a supply disruption in one country would not greatly affect world supplies, although there might be a sharp rise in the world price for a short period because of short-run inelasticity of demand. Only in the case of metals such as chromium, cobalt, manganese and columbium could production be sufficiently affected by a supply disruption in one producing country to create a serious world shortage.

In the case of the USA, supplies of strategic minerals could be made available from the national defense stockpile for use by defense industries, but not for civilian consumption (Mikesell, 1986a). A few other countries maintain limited government stockpiles of nonfuel minerals and the US government has made some effort to encourage stockpiling by the governments of Japan and the European countries. There has also been some interest in the USA in the creation of an economic stockpile, materials from which could be sold on the market in the event of a peacetime supply disruption. However, the cost effectiveness of an economic stockpile is debatable in the light of the relatively low probability of a substantial disruption of supply of any nonfuel mineral that would last more than a few months (Mikesell, 1985).

Industrial countries could reduce their vulnerability to supply disruptions by cooperative action similar to the International Energy Agency program for dealing with potential disruptions in petroleum supplies. Each Western industrial country could assume responsibility for accumulating agreed inventories of those nonfuel minerals subject to supply disruption. In addition, each cooperating nation would agree not to impose export restrictions or other interferences on the stockpiled minerals in trade with other cooperating members. Finally, there should exist coordinated programs for stockpile releases in the event of a global supply disruption. Stockpile releases would take place only in response to designated types of disruptions in the source countries resulting in a specific percentage reduction in world supply or a specific percentage increase in world price. The nonfuel minerals selected for inclusion in the program should be limited to those for which there exists significant probability of substantial supply disruption over the next two or three decades.

Seabed Nodules and the Law of the Sea Treaty

Seabed or polymetallic nodules were discovered in the Pacific by the British research vessel *HMS Challenger* in 1876, but remained a scientific curiosity until the mid-1960s when several mining companies began to locate concentrations and develop technology for recovery. The principal metals in the nodules are manganese, nickel, copper and cobalt, but metal content differs from area to area. In the north and central Pacific, where initial ventures are likely to take place, an extraction unit with a capacity of 3 million mt of nodules per year would yield about 35,000 mt nickel, 30,000 mt copper, 5,000 mt cobalt and 600,000 mt manganese. These quantities would be sufficient to satisfy approximately 5 per cent of annual (in 1979) world demand for nickel; about 22 per cent of the demand for cobalt; about 30 per cent of the demand for manganese; and about 0.3 per cent of the demand for copper (Ministry of Natural Resources, 1980, p. 17). Since it has been estimated that each extraction unit should have a capacity of at least this amount for economical recovery, the output of three such units would satisfy the bulk of world demand for cobalt and manganese and a significant portion of the world demand for nickel. This output would, of course, be in direct competition with land-based output so that countries producing these minerals have been concerned with the possibility that world markets might be swamped by minerals from the sea.

Mining operations for collecting and processing seabed nodules will be very costly, requiring perhaps a minimum of $2 billion (in 1985 dollars) for each operation. Also, there remain a number of technical problems involving both mining and metallurgy that must be solved before commercial operations can begin. Four consortia, each formed of mining companies of different nationalities, and one French consortium have been active in exploration and research. The consortia and their members are listed in Table 7.2. By 1982 an estimated $400 million had already been spent on exploration and research by the consortia, but several factors have resulted in a sharp decrease in their activities since 1980. First, and perhaps foremost, are the uncertainties regarding the outcome of the Law of the Sea (LOS) conference and establishment of the proposed International Seabed Authority (ISA). A second factor has been the sharp decline in prices of the four principal minerals in seabed nodules. Much will depend upon the future price of nickel since an estimated two-thirds of the revenue from seabed production will be derived from nickel.

An important factor determining the feasibility of seabed nodule production is the cost relative to that for land-based output. Under current conditions, seabed mining would not have a cost advantage over land-based mining, but it is expected that as high-grade land-

Table 7.2   *Major Seabed Mining Consortia*

Participants	Parent company	Country of origin
	Kennecott Consortium	
Kennecott Corp.	Sohio	USA
RTZ Deepsea Enterprises	Rio Tinto Zinc	UK
Consolidated Gold Fields	Rio Tinto Zinc	UK
BP Petroleum Development	British Petroleum	UK
Noranda Exploration	Noranda Mines	Canada
Mitsubishi Group	Mitsubishi Corp.	Japan
Mitsubishi Metal		
Mitsubishi Heavy Industries		
	Ocean Mining Associates	
Essex Minerals	US Steel Corp.	USA
Union Seas	Union Minière S.A.	Belgium
Sun Ocean Ventures	Sun Company	USA
Samim Ocean	Ente Nazionale Idrocarburi	Italy
	Ocean Management Incorporated	
INCO	INCO	Canada
AMR	Metallgesellschaft	West Germany
	Preussag	West Germany
	Salzgitter	West Germany
SEDCO		USA
Deep Ocean Mining Co.	23 companies	Japan
	Ocean Minerals Company (OMCO)	
Amoco Ocean Minerals	Standard Oil of Indiana	USA
Lockheed Systems Co.	Lockheed Aircraft Corp.	USA
Ocean Minerals Inc.	Billiton B.V.	Netherlands
	Association française pour l'étude et la recherche des nodules (AFERNOD)	
Centre national pour l'exploitation des océans (CNEXO)	Public corporation	France
Commissariat à l'énergie atomique (CEA)	National centre for atomic energy	France
Société métallurgique le nickel (SLN)		France
Chantiers de France-Dunkerque		France

Source: Department of International Economic and Social Affairs, *Sea-Bed Mineral Resource Development*, New York: United Nations, 1982.

based reserves decline, the relative cost position of seabed mining will improve. For example, sulphidic nickel ores are cheaper to produce than lateritic ores, but as demand for nickel expands a larger portion will be met from more abundant lateritic nickel mines. Hence, it is customary to compare seabed mining costs of producing nickel with the cost of lateritic nickel mining. Lower cost technologies are being developed to process lateritic ores, but there are also substantial opportunities for economies of scale from deep sea mining once technologies for exploration, extraction and processing have been standardized (Dick, 1985, Chapter 1). It is difficult to project the relative cost of sea-bed mining compared to land-based mining. Because of the huge investment required, it seems unlikely that private seabed mining will be initiated until there is a substantial rise in real prices of metals and the costs of producing metals from the seabed fall significantly lower than those for new land-based mines.

In addition to deep sea nodules there are large amounts of minerals contained in crustal formations on the continental shelf of North America and other continents. Exploration is just beginning and relatively little is known about their economic potential. Because crustal formations are not as deep as seabed nodules, some believe they may be more economic to mine. Since these resources are within the 200-mile limit, exploiting them would not create the international political problems associated with exploitation of seabed nodules.

Seabed Minerals and the United Nations Law of the Sea Convention

The fact that seabed nodules are found thousands of miles beyond the continental shelf of any country has raised the question of ownership and the right to exploit them. The fundamental position of the USA and certain other industrial countries is that seabed mining is a freedom of the high seas and that such activity should be available to private citizens of any country as well as governments. However, just as shipping, fishing and air transport over the oceans are subject to international regulation, so also should be seabed mining. Regulations are necessary to avoid conflicts over claims to the areas mined based on discovery, and to prevent pollution and interferences with other activities. Industrial countries recognized that seabed mining was a proper subject for consideration by the series of UN Law of the Sea (LOS) conferences that began in the late 1950s.

The UN LOS conferences have been concerned with dozens of issues, including commercial navigation through straits, fishing and pollution. The first and second LOS conferences met in 1958 and

1960 respectively, but serious negotiations on conditions for mining seabed nodules took place during the third LOS conference, which began in December 1973 and continued through eleven sessions to December 1982. Contrary to the position of most industrial countries, Third World countries participating in the conferences argued that seabed nodules constitute a "common heritage of mankind" and therefore should be exploited solely by an International Seabed Authority (ISA) for the benefit of all nations and not be available to private interests. After long negotiations a compromise was embodied in the draft LOS Convention approved on April 30, 1982 by a majority of the nations represented, with 130 nations voting in favour of the convention. However, four nations – the US, Israel, Turkey and Venezuela – voted against the convention; and there were seventeen abstentions, including the major Western European countries (except France and the Scandinavian countries) and most Soviet Bloc countries. At a subsequent meeting held on December 10, 1982, 117 nations signed the draft convention. Signatories included most developing countries, the Soviet Bloc countries and all developed countries except Belgium, West Germany, Italy, the UK, the USA and Japan. Once sixty nations ratify the treaty, it will go into operation. Since most of the capital and enterprise for development of seabed nodules must come from the industrial countries that oppose the convention as now drafted, little progress is likely to be made toward their development.

Provisions of the LOS Convention Relating to Seabed Mining
The LOS Convention[3] provides for establishment of an ISA with broad powers over seabed mining, including approval of mining contracts with private or national governmental enterprises. Companies desiring to mine must explore two sites and make the information on both sites available to the ISA, which at its option has the right to develop one of the sites. The mining company may then obtain a contract to develop the other site under certain conditions and provided the contract is approved by majority vote of the Technical Commission of the ISA. The Technical Commission is appointed by a three-fourths majority of the Executive Council made up of representatives from thirty-six nations operating under a one-nation, one-vote rule. Given the high proportion of Third World member countries, it is feared that industrial countries are likely to have little voice in ISA decisions.

Following the ratification of the LOS Treaty, mining activities by the ISA are to be undertaken by a company called the Enterprise, which is empowered to mine sites explored by private or national companies without cost to the Enterprise. The Enterprise is also guaranteed access to seabed mining technology owned by private companies applying for contracts, together with the technology used by these companies but owned by others. Similar access to privately owned technology is provided to developing countries planning to go into seabed mining. Such technology must be sold on reasonable terms to the Enterprise as a condition for granting contracts to private or governmental entities. Financing for operations of the Enterprise is to be provided by interest-free loans and assessments from treaty members on the scale used for regular contributions to the United Nations budget. (The US share of such contributions has been 25 per cent.)

Firms receiving mining contracts must pay an application fee and an annual fee on the contracts; with the beginning of commercial production contractors must pay either a production charge ranging from 5 to 12 per cent of the gross value of output, or a lower production charge (2 to 4 per cent) plus a share of net profits. This levy ranges from 35 to 50 per cent of net profits during the first ten years and 40 to 70 per cent during the second ten years.[4] The percentage of net profits paid to ISA is graduated on the basis of net profits in excess of specified rates of return on the contractor's investment. Revenues from seabed mining, including those from operations of the Enterprise and the levies paid to ISA by contractors, are to be divided among the 150 nations expected to ratify the treaty. The treaty also provides for shares of revenues to be paid to national liberation movements such as the Palestine Liberation Organization (PLO).

The USA and other industrial countries objected strongly to the initial draft of the LOS Convention since it gave no assurance that the international consortia listed in Table 7.2, which have spent large sums in pioneering exploration and research on seabed mining, would be able to obtain contracts. The April 1982 draft convention, therefore, guarantees at least one seabed mining contract to each of these international consortia and to three state-sponsored programs by the Soviet Union, India and China.

In order to protect the interests of countries producing land-based minerals, the convention provides a limit on the production of seabed minerals. There is also provision for international buffer stocks to stabilize prices of seabed minerals and for compensation to countries adversely affected by decreases in the prices received for land-based output.

After fifteen years of production, the convention provides that treaty provisions are to be reviewed to determine whether the policy objectives have been fulfilled. If two-thirds of the member states decide to amend the treaty they may do so after five years of negotiations and ratification by three-fourths of the member states.

Efforts by Industrial Countries to Amend the Draft Convention
During the eleventh session of the LOS Conference (March–April 1982), seven industrial countries having a special interest in seabed mining – Belgium, France, Germany, Italy, Japan, the UK and the USA – put forth a number of amendments to the mining provisions of the LOS Treaty, the most important of which were:

(1) a guarantee that the pioneer seabed investors could obtain mining contracts
(2) an increase in the majority of the Executive Council needed for certain key decisions
(3) a provision that no amendments to the seabed provisions would take effect unless ratified by all parties to the treaty or would not bind the states opposing the amendments
(4) a requirement that the Enterprise obtain mine sites for its own use on the basis of a random selection rather than by choosing the better of two sites proposed by each applicant
(5) the guarantee of a seat on the Executive Council for the USA
(6) a modification of the requirements on contractors to transfer technology to the ISA.

Most of the proposed amendments were rejected by a majority of the delegates to the conference and only the amendment guaranteeing a control to pioneer seabed investors was adopted.

Position of the Reagan Administration on the LOS Draft Convention
During the Ford and Carter administrations, the US delegation to the LOS Conference negotiated several compromises with Third World delegates whose initial position was that seabed mining would be conducted exclusively by the ISA, either directly by the Enterprise or by some association between the Enterprise and private corporations or states. From early 1977 to October 1980 the US delegation was headed by Ambassador Elliot L. Richardson, whose position was that while many provisions relating to seabed mining were flawed, the USA had much to gain from those provisions of the convention concerned with navigation, fisheries, pollution and other matters unrelated to mining. However, at the beginning of the tenth session in March 1981, the Reagan administration announced that it was reviewing the entire draft convention and US policy with respect to it, and the US delegation would not participate fully in the conference until that review had been completed. On January 29, 1982 President Reagan announced that the USA would seek fundamental changes in the draft convention at the eleventh session scheduled for March 1982.[5] The President stated that the treaty must not deter development of any deep seabed mineral resources for meeting national and

world demands. Among the provisions of the draft treaty that discouraged development of these resources were: (1) the policies of the ISA that provide for a production ceiling limiting the availability of minerals for global consumption; (2) a limit on the number of mining operations that could be conducted by any one country, thus potentially limiting the ability of US firms to supply US consumption needs; and (3) the broad areas of administrative and regulatory discretion which, if implemented in accordance with ISA production policies, would deter seabed mineral development. The President also stated that the treaty must assure access to those resources by US nationals to provide US security of supply, avoid monopolization of the resources by the ISA and promote the economic development of the resources. In this regard, the administration maintained that the rights of private companies that made pioneer investments in seabed mining were not fully protected and that the draft convention created a system of privileges that discriminated against private operations.

The President was also dissatisfied with the role of the USA in the ISA decision-making process based on one nation, one vote. He objected to the provisions for mandatory transfer of private technology and to participation in the revenues of the ISA by national liberation movements. Finally, the President objected to the provision that two-thirds of the member states could amend the treaty in a way that was inimical to US interests. (The two-thirds vote was later changed to a three-fourths majority vote.)

The positions taken by President Reagan were in line with those adopted by Congress in the Deep Seabed Hard Minerals Resources Act of June 28, 1980 which set forth conditions for any international agreement to which the USA would become a party.[6] These include "assured and nondiscriminatory access under reasonable terms and conditions, to the hard mineral resources of the deep seabed for US citizens." The Act also provides for US government licensing of exploration and mining of seabed minerals until such time as the USA becomes a party to an international treaty governing seabed mining. The Act provides for environmental regulations and a royalty payment of 3.75 per cent of the market value of the minerals, with the proceeds to be placed in a fund that might be used for payments to an international authority set up by international treaty. In addition, the Act provides for the designation of any foreign nation as a "reciprocating state," if that nation regulates the conduct of its citizens in a manner compatible with the provisions of the Act. Special treaties might be negotiated with such states to regulate exploration and commercial recovery of seabed minerals in a manner designed to avoid conflict over claims. After adoption of the draft convention in 1982, the USA negotiated an agreement with France,

West Germany and the UK to resolve any conflicting claims filed by seabed mining consortia. However, this so-called "mini treaty" would not be valid in the event any of the signatories ratified the LOS Treaty.

President Reagan's conditions for acceptance of the LOS Convention were so fundamentally at odds with the draft convention that the vast majority of delegates was unwilling to seriously debate them. It has been argued by Richardson and others that had the USA participated fully in the tenth session, there might have been a better chance of modifying the convention before it was voted on in April 1982 (Richardson, 1983, pp. 505–17). However, in the light of the rejection by the majority of the delegates of amendments to the draft convention put forward by the seven industrial countries during the eleventh session prior to the final vote on the convention, it is doubtful whether any of the more fundamental conditions set forth by President Reagan would have been acceptable to the majority of the delegates to the conference.

Public Debate over US Ratification of the LOS Treaty
The refusal of the Reagan administration to sign the LOS Treaty resulted in widespread debate in the USA and in other industrial countries that followed the same course. The most knowledgeable and articulate critic of the US action was Ambassador Richardson who argued that despite its flaws, the USA should either have voted for the treaty or abstained, as did most European countries. Richardson believed that abstention would have made possible further efforts to negotiate changes in the convention before final ratification. But even if the USA were not able to make significant changes in the draft convention, Richardson believes the deficiencies from the standpoint of the USA were not so great as to make the LOS Treaty unworkable and that the advantages of other provisions may be lost if the USA is not a signatory to it. Finally, Richardson argued that in the absence of US membership in the ISA, US mining firms would have no legal right to exploit sites in the seabed and that in the event of dispute the International Court of Justice would declare US seabed mining operations illegal (Richardson, 1982, pp. 89–110).

In opposition to Richardson's position, James L. Malone, Assistant Secretary of State and President Reagan's special representative to the LOS Conference, argued that the USA would lose nothing by not ratifying the treaty since the nonseabed mining provisions would be available to the USA even if it did not sign. Malone also argued that the provisions requiring sale of technology to other countries would contravene US law and the Senate would never ratify the treaty.

Malone further argued that no US mining firm would engage in seabed mining under the terms of the treaty, but might do so under US law.

The American public and press have been deeply divided over the issue of signing the LOS Treaty. The American Mining Congress and representatives of the American consortia undertaking exploration of the seabed have been strongly opposed to the treaty and have indicated they could not operate under its financial and other provisions. The *Wall Street Journal* was adamantly opposed to the treaty, while the *New York Times* and *Washington Post* were more favorable. Support for and against US adherence to the treaty has not been neatly divided among liberals and conservatives. An article in *Fortune* (Alexander, 1982, pp. 129–44) tended to favor the treaty while the *New Republic* printed an article strongly opposing US ratification.[7]

Conclusions on the Effects of the LOS Treaty on Development of Seabed Minerals

The adoption of the LOS Treaty has undoubtedly delayed development of seabed minerals, and this would be true whether or not the US signed. Fortunately, the world is not likely to need this source of minerals until the next century, by which time the international political climate may have shifted to allow the treaty to be amended in a manner that will foster seabed mining by both private and government enterprises. A sharp rise in the prices of minerals could be a potent force for overcoming political obstacles to investment, but without a strong price incentive seabed mining is unlikely to attract the large amounts of required capital.

International Assistance for Mineral Development in Third World Countries

Beginning in the mid-1970s there was substantial interest on the part of development assistance agencies in promoting mineral production in Third World countries. The importance of developing petroleum and other energy resources in these countries was obvious since the oil-importing LDCs were severely affected by the several-fold rise in oil prices during the 1970s. The World Bank's interest in nonfuel mineral development arose from several factors. First, Third World countries were strongly advocating, both directly and through UNCTAD, the establishment of an international trust fund designed to finance government mineral enterprises. Most Third World countries favored government enterprise over foreign private investment for development of their mineral resources and, therefore, sought public international financing for government petroleum and mining projects. A related factor was the decline in foreign private investment in Third World mining as a consequence of expropriations and poor climates for foreign investment. A third factor that may have influenced the World Bank was the widespread fear of a world shortage of nonfuel mineral producing capacity.

The USA and other industrial countries were not averse to international development financing institutions becoming more active in promoting mineral development in the LDCs. However, most industrial countries wanted international institutions to use their resources as a catalyst for promoting foreign private investment rather than for financing state enterprises. For example, the International Finance Corporation (IFC), an affiliate of the World Bank, can take a minority equity position or provide a portion of the debt financing for foreign or domestically owned private mining companies in Third World countries. Participation by the IFC provides confidence to potential private investors since governments are unlikely to expropriate or violate investment contracts in projects in which the World Bank Group has an interest. Also, either the IFC or the World Bank can make loans to joint ventures involving state mining enterprises and foreign investors in a way that makes investing more attractive to the foreign investors.

The divergence of interests between Third World and industrial countries over the method of financing mineral resource development is well illustrated by the debate over the proposal by the then Secretary of State Henry Kissinger for an International Resources Bank (IRB) at the 1976 UNCTAD meeting. The proposed IRB was designed to guarantee and provide loan financing to foreign private investment in resources. The IRB proposal was rejected by a majority of Third World countries who favored World Bank Group loans to SMEs for exploration and development of minerals.

By and large the US position with respect to World Bank and Inter-American Development Bank (IADB) loans to SMEs has not prevailed. Between mid-1978 and mid-1985 World Bank and International Development Association loans for mining totaled nearly $600 million, most of it to SMEs (World Bank, 1983, 1984 and 1985). The IADB made several loans to the Peruvian government enterprise, Centromin (a mining complex formerly owned by American investors under the name Cerro de Pasco before the company was expropriated in 1974). A loan of $268 million was made in 1983 by the IADB to the Chilean SME, CODELCO, to expand copper production. This loan was highly controversial because of excess world copper producing capacity.[8] During the Reagan administration the US executive director on the board of the World Bank has voted against a number of loans to SMEs, but the USA does not have a veto over such loans. The US executive director in the IADB also voted against loans to SMEs. Smaller amounts of loan and equity participation in private mining companies in developing countries have been made by the IFC, and the US government has not only favored such loans but helped to promote a recent increase in the capital resources of the IFC. During the calendar years 1978–84 the IFC made loans and equity investments in metal mining and processing projects totaling about $300 million (International Finance Corporation, Annual Reports, various issues).

The position of the US government on international public agency lending to Third World SMEs raises serious questions of foreign economic policy. The arguments presented by the American mining industry and Congressmen supporting them are that such loans subsidize foreign competition with the USA and, in addition, encourage worldwide overcapacity in mining. But there is also world overcapacity in steel, textiles, footwear and a number of agricultural commodities. If the US government adopted a policy of preventing development loan assistance to any Third World industry that competed with US firms in world markets, the scope for financing economic development would be greatly reduced. An effective development assistance program must include promotion of export industries in developing countries and a high proportion of these industries compete with the production of developed countries.

One argument against international public agency loans to SMEs is that since private international capital is available for financing Third World mining activities – provided a favorable investment climate exists – these agencies should conserve their resources for financing projects for which foreign private capital is not available. This argument is in accordance with provision in the World Bank charter that states the bank's loans should be available to *supplement* private international financing, not to displace it. Third World countries argue that discrimination against SMEs violates the provision of the bank charter that forbids discrimination against countries on the basis of their political and economic structure. Hence, if a country's political orientation favors SMEs over foreign private enterprise for development of resources, this should be respected by the bank. The apparent contradiction between these provisions in the bank charter has never been resolved. It would appear, however, that the bank should evaluate loan applications on the basis of both the viability of the project, including world market conditions, and the managerial and technical competence of the borrower, without regard to whether the borrower is a private or government enterprise.

Should the World Bank and other international development agencies make loans in support of projects when there is world overcapacity? Some loans have been made to SMEs to enable them to maintain capacity or to reduce costs and increase productivity rather than expand capacity. We may question whether international public agencies should be guided by judgments regarding the existence of overcapacity in any industry. In a competitive world economy there is frequently world overcapacity in an industry, but firms with lower costs tend to survive and expand while those with higher costs must adjust or disappear. In considering a loan, the bank should be concerned with the viability of the project rather than the impact of output on world prices and competitive conditions.

Finally, there is the argument that international public agency loans subsidize Third World production in competition with private firms in developed countries. World Bank and IADB loans are made at interest rates more favorable than those obtained by private firms in the international financial markets. It might well be argued that a very efficient copper mining firm, such as Chile's CODELCO, should obtain its financing from private sources, while in view of its limited loan resources the World Bank ought to concentrate on loans for infrastructure, peasant agriculture and other domestically consumed production. This is perhaps the soundest argument against making loans to Third World country enterprises engaged in production for world markets. However, the decisions on loans should be on a case-by-case basis, taking into account both the viability of the project and the optimum allocation of the loan institution's resources rather than on the basis of general rules, such as the elimination of loans to SMEs or to particular industries that may compete with industries in developed countries.

Notes

1 The USA became a member of the International Coffee Agreement in 1964. It joined the fifth ITA in 1976, but in October 1982 the Reagan administration announced the USA would not join the sixth ITA, mainly because the principal tin producing members employed export quotas to support tin prices rather than relying solely on the operation of the buffer stock.

2 For presentation of opposing views by economists specializing in commodity markets, see Adams and Klein (1978). The papers in the volume were presented at a Conference on Stabilizing World Commodity Markets sponsored by the Ford Foundation held at Airlie, Virginia, March 1977.

3 For the text of the LOS Convention see United Nations (1982). The analysis of the convention in this section was based on a number of journal articles, including Zuleta (1983) and Larsen (1982).

4 During the first ten years the share to be paid to the ISA is 35 per cent on that

portion representing a return on the investment of less than 10 per cent; 42.5 per cent on a return between 10 and 20 per cent; and 50 per cent on a return over 20 per cent. In the second ten years of commercial production the corresponding shares to be paid to the ISA are 40 per cent, 50 per cent and 70 per cent respectively. United Nations (1982, Article 13).

5 For the text of President Reagan's statement see US Department of State (1982, pp. 54 ff). For an analysis of that statement, see Malone (1982, pp. 61–3).
6 Public Law 96–283, 96th Congress, 1st session, June 28, 1980.
7 Chapman (1982, pp. 17–20). For another criticism of the LOS Treaty, see Burke and Brokaw (1982, pp. 71–82).
8 This estimate was derived from World Bank *Annual Reports* for 1978–85.

8

The Future of the World Mining Industry

Introduction

Consideration of the future of the world mining industry involves several time frames and categories of physical, technological, economic and organizational evolution. In the very long run, say, the next half-century, we are concerned with the outlook for world mineral resources in relation to the global demand for them. Over the next quarter-century we are concerned with structural changes in the world mining industry. There will be shifts in the geographical concentration of output for various minerals, and changes in mining and processing technology and in financial and organizational structure. Finally, in considering the next decade we are concerned with changes in real prices of minerals, with shifts in world markets and competitive positions of mining firms in various countries, and with the level and pattern of investment in mining.

Will the World Run Out of Nonfuel Mineral Resources?

An important debate among resource economists over the past several decades concerns whether and when world economic progress will be halted or significantly retarded by the exhaustion of mineral resources. The resource pessimists point to growing world population and rising per capita consumption in the face of limited recoverable resources in the earth's crust, while the resource optimists point to advances in technology for discovering and extracting additional reserves and substituting more abundant minerals for those that may become scarce. We cannot predict with confidence either world demand for nonfuel minerals or the additional reserves that will be discovered over the next half-century. Nor can we predict the degree to which technological progress will enable us economically to mine lower ore grades or to substitute minerals derived from abundant resources, such as clay, for those now extracted from mines. We can, however, extrapolate trends in resource availabilities and requirements, in technological progress in mining and metallurgy, and in conservation of nonfuel minerals over the past half-century. Our analysis of these trends provides a basis for optimism regarding the future availability of nonfuel minerals at a cost that will not impair world economic progress.

Several developments over the past twenty-five years have had an important bearing on the outlook for nonfuel mineral supplies in relation to demand well into the twenty-first century. First, despite the rapid growth of world industrial production, nonfuel minerals remain adundant and none of the major minerals can be said to have become increasingly scarce. Increasing or decreasing scarcity of a mineral may be measured by long-run trends in either its real price or its real cost of production. Although mineral prices exhibit cyclical

Table 8.1 *Weighted Price Index of Ten Metals and Minerals in Constant Dollars[a] (1977–9 = 100)*

1950	125	1970	143
1951	135	1971	117
1952	150	1972	107
1953	139	1973	135
1954	136	1974	149
1955	156	1975	116
1956	159	1976	111
1957	139	1977	104
1958	127	1978	92
1959	125	1979	104
1960	125	1980	106
1961	120	1981	93
1962	115	1982	85
1963	115	1983	88
1964	138	1984	85
1965	158		
1966	164	*projected*	
1967	136	1985	84
1968	140	1990	85
1969	146	1995	90

[a] Weighted by 1977–9 developing countries' export values

Note: Commodities: copper, tin, nickel, bauxite, aluminum, iron ore, manganese ore, lead, zinc and phosphate rock

Source: Economic Analysis and Projections Department, *Half-Yearly Revision of Commodity Price Forecasts and Quarterly Review of Commodity Markets for June 1985*, Washington, DC: World Bank, July 1985, table 3, p. 12.

Table 8.2 *Indexes of Commodity Prices and Price Projections in 1981 Constant Dollars*

	Average 1960–70	1976	1980	1981	1982	1983	Projected 1995
Copper	201	118	120	100	86	96	103
Tin	73	79	113	100	92	96	89
Nickel[a]	85	97	95	100	95	98	77
Aluminum[b]	111	86	87	100	102	107	106
Lead	109	90	119	100	77	61	79
Zinc	104	124	86	100	90	95	112
Iron ore	209	132	101	100	108	103	95
Bauxite	79	100	98	100	92	91	92
Manganese ore	146	127	89	100	100	95	87

[a] Canadian producer price
[b] US producer list price
Source: Indexes were calculated from price data in World Bank, *Price Prospects for Major Primary Commodities*, Vol. IV: *Metals and Minerals*, Washington, DC: World Bank, September 1984.

fluctuations, over the longer run real prices will tend to reflect the real marginal cost of finding and extracting new reserves. As is shown in Table 8.1, the weighted average index of prices of ten major minerals and metals in constant dollars has trended downward since the mid-1960s, and, according to a 1985 World Bank study, the projected index for 1995 is substantially lower than the index for every year during the period 1950–77. Moreover, projected 1995 prices in constant dollars for six of nine major nonfuel minerals are less than their average prices during the 1960–70 period (see Table 8.2). These figures suggest that average real prices of major nonfuel minerals as a group will be no higher at the end of this century than they were for the 1960–70 period.

A second factor relevant for judging the availability of nonfuel minerals in future decades is the trend in mineral reserves and potentially recoverable mineral resources. Measuring reserves of known orebodies requires sufficient collecting of core samples to estimate the volume and grade of the ore with a high degree of confidence. Measured reserves include only minerals that can be economically mined at current prices and costs; they constitute a relatively small portion of potentially recoverable minerals. Indicated and inferred reserves are determined on the basis of less intensive examination or are inferred from geological evidence. In addition to presently economic reserves, there are marginal reserves and subeconomic resources that may potentially become reserves (see

Figure 8.1 *Major Elements of Mineral Resource Classification*

Identified resources   Undiscovered resources

Measured	Indicated	Inferred	Hypothetical	Speculative
Reserves		Inferred reserves		
Marginal reserves		Inferred marginal reserves		
Subeconomic resources		Inferred subeconomic resources		

Source: Bureau of Mines, *Mineral Commodity Summaries 1982*, Washington, DC: US Department of Interior, 1982.

Figure 8.1). The "reserve base" of a mineral includes measured reserves, marginal reserves and that portion of subeconomic reserves expected to become economic within a few decades.

All the classifications mentioned above are included in *identified* resources, but there are two classifications of *undiscovered* resources that potentially may become economical to exploit: (1) *hypothetical* resources in known mining districts, and (2) *speculative* resources in nonproducing districts, but in favorable geologic settings (Figure 8.1). Hypothetical and speculative resources are defined as potentially commercial, but the vast bulk of the undiscovered resources lie outside both these categories. As more geological information is obtained, more resources will be added to one of the classifications listed in Figure 8.1.

One method of measuring the future availability of a mineral is to calculate the number of years before the reserves of that mineral will be exhausted, given the expected annual rate of increase in consumption. For most major minerals, the life expectancy of the 1980 reserve base (which includes currently economic reserves plus some resources with a reasonable potential for becoming economic over the next several decades) ranges between twenty-five and fifty years. There are, however, several weaknesses with this measure. First, new discoveries continually increase reserves and reserves of nearly all minerals have been growing despite the fact they are being depleted by production. In addition, the life expectancy of reserves of nearly all major metals has tended to increase over the past couple of decades due to reduced rates of consumption; in some cases these rates are expected to decline further in the future. Second, it is difficult to project rates of growth in consumption beyond a decade or so since these rates are determined by technological developments, patterns of world consumption and substitution in the use of minerals brought about by changes in their relative prices.

A more realistic approach to determining future availability is to use mineral resources rather than reserves. The classification of mineral resources used by the BOM includes all identified resources plus, in some cases, hypothetical and speculative deposits. All mineral resources are in such forms and amounts that economic extraction is currently or potentially feasible.

Table 8.3 gives world resources for twenty-two minerals. In most cases the estimates of mineral resources include all identified resources, and some include hypothetical and speculative resources, such as minerals contained in seabed nodules. Table 8.4 gives the life expectancy beyond 1980 for these resources at projected annual rates of growth in consumption estimated by the BOM for the 1978–2000 period. It will be observed that for seven of the twenty-two minerals the expected life exceeds 100 years; for another eight the resources would be sufficient to meet consumption requirements to the year 2050; and, except for industrial diamonds and gold, resources for the remainder would be sufficient to meet requirements to the year 2025. Given the likelihood that world resources will continue to expand, this should make us feel comfortable regarding availability of supplies to meet world demands for most major minerals through the middle of the next century – except for the fact that we have little basis for projecting the rates of growth in consumption of minerals during the twenty-first century.

In 1980 the industrial countries, including the Soviet Bloc, consumed about 80 per cent of total nonfuel minerals produced, but these countries accounted for less than 25 per cent of world population. In the next century the developing countries' demand for nonfuel minerals is expected to grow faster than that of the developed countries. In the coming decades both GNP and population growth rates are likely to be higher in developing than in developed countries. This has been the case since 1960. Most developed countries will have zero population growth by early in the next century, while most developing countries will not reach stationary population until late in the twenty-first or in the twenty-second century. Furthermore, the composition of consumption in developing countries will tend to favor goods containing relatively higher amounts of nonfuel minerals, while the consumption in developed countries will tend to favor services requiring much lower inputs of minerals. These factors provide the basis for the pessimistic

Table 8.3 *World Resources of Selected Minerals*[a]

Mineral	Volume
Bauxite	55–75 billion mt
Chromium	36 billion st gross weight chromite
Cobalt (metal)	12 million st land + 250 million st sea
Columbium (metal)	38 billion lb
Copper (metal)	1,627 billion mt land + 0.76 billion mt sea
Diamonds (industrial)	1+ billion carats[b]
Gold (metal)	2.4 billion oz, of which 15–20% are byproduct resources
Ilmenite	1 billion st of contained titanium dioxide
Iron ore	260+ billion st of Fe
Lead (metal)	1.4 billion mt
Manganese (metal)	3.1 billion st land + 15.0 billion st sea
Mercury	17 million 76–lb flasks
Molybdenum	46 billion lb
Nickel	143 million st land + 430 million st sea
Platinum group metals	3.3 billion troy oz
Rutile	200 million st of contained titanium dioxide
Silver (metal)	25 billion troy oz
Tantalum (metal)	560 million st
Tin (metal)	37 million mt
Tungsten (metal)	14.9 billion lb
Vanadium (metal)	140+ billion lb
Zinc (metal)	1.8 billion mt identified + 2.5–3.0 billion mt, hypothetical and speculative

[a] Mineral resources are in such forms and amounts that economic extraction is currently or potentially feasible. The classification of resources applying to each of the minerals listed is not uniform. In most cases they include all identified resources, but may also include hypothetical and speculative deposits (see Figure 8.1).
[b] Includes newly discovered Australian reserves estimated as high as 500 million carats.
Note: mt = metric ton; st = short ton
Source: Bureau of Mines, *Mineral Commodity Summaries 1986*, Washington, DC: US Department of Interior, 1986; and Bureau of Mines, *Mineral Facts and Problems, 1980*, Washington, DC: US Department of Interior, 1981.

prediction that economic growth, particularly in the developing countries, will be halted or seriously constrained by a scarcity of nonfuel minerals, perhaps well before the middle of the next century.

On the basis of world population projections and recoverable resources estimated by the US Geological Survey (USGS), F. E. Trainer estimates that if each individual in the world in 2050 were to consume as much as each American did in the 1970s "the recoverable resources of eleven of the twenty-four most common minerals would

Table 8.4 *Life Expectancy beyond 1980 of World Resources of Selected Minerals at Probable Average Annual Rates of Growth in Consumption for the Period 1978–2000 (in years)*

Bauxite	74
Chromium	Over 100
Cobalt	Over 100
Columbium	70
Copper	75
Diamonds (industrial)	20
Gold	35
Ilmenite	48
Iron ore	80
Lead	85
Manganese	Over 100
Mercury	50
Molybdenum	52
Nickel	Over 100
Platinum group	Over 100
Rutile	70
Silver	45
Tantalum	80
Tin	100
Tungsten	49
Vanadium	Over 100
Zinc	Over 100

Source: See Table 8.3 for estimates of world resources. Includes seabed resources for cobalt, copper, manganese and nickel. For probable average annual growth rates for the period 1978–2000 see Bureau of Mines, *Mineral Facts and Problems, 1980*, Washington, DC: US Department of Interior, 1981. Annual rates of growth in consumption refer to growth demand for primary metal. A growing portion of the demand for the metal is expected to be supplied by secondary or recyled metals.

be exhausted before 2050" (Trainer, 1982, p. 50). Trainer concludes that "if the standard of living of the West is to be extended to the anticipated world population of the mid-twenty-first century, something like seven or nine times as much of each mineral would have to be discovered each year as has been found annually in the last few decades, and this would have to be repeated for as long as an affluent world society is to keep running."[1] Such a rate of discovery is most unlikely.

If developing countries can only achieve the current real per capita GNP of developed countries by consuming the same amount of minerals per person as is currently consumed in developed countries, there is virtually no chance that the developing countries can enjoy the same real per capita consumption currently existing in the developed countries. Barring some technological breakthrough that cannot presently be perceived, there simply are not enough recoverable minerals to support this level of affluence on a global basis. This suggests that developing countries can only hope to expand food supplies and shelter and eliminate abject poverty; they must forget about two cars in every garage and a house or apartment filled with electric appliances.

However, this pessimistic outcome is by no means inevitable. There are some minerals, such as aluminum, iron and silicon, that are abundant in the earth's crust beyond any possible level of utilization by mankind, and these minerals can conceivably serve as substitutes for less abundant materials, For example, more than 8 per cent of the earth's crust consists of aluminum, the most abundant metal; iron takes second place at 5 per cent. Modern technology has already led to the substitution of fiber optics (produced from sand) for copper, and ceramic materials (produced from clay) for iron and other metals. Materials technology has been advancing very rapidly in response to supply limitations signaled by rising prices for individual minerals. Moreover, the potential for recycling and conservation of less abundant minerals is enormous. It may be that technology and energy are the only true factors limiting growth, and by the middle of the next century we may have learned to harness the almost unlimited energy resources of the sun. We simply do not know enough to forecast when if ever the availability of nonfuel minerals will constitute a limit to world economic progress.

Structural Changes in the Mining Industry: The Next Fifteen Years

There have been important trends in the world mining industry over the past fifteen years that seem likely to continue through the end of this century and beyond. These trends have reversed the popular perception that nonfuel mineral resources are becoming scarce. They are in large measure a product of the technological revolution in materials that is rapidly changing the industrial structure of the world. Their impact on the world mining industry is likely to be as radical as the changes in the automotive and electronic industries over the past decade. The following paragraphs summarize these long-term trends and their probable implications for the mining business and international trade in minerals.

(1) The rates of growth in world demand for major metals have declined by more than half since the mid-1970s. Abstracting from the 1982–3 recession, US and EEC demand for most metals has been flat

or declining since 1978. Slower growth in metals consumption combined with increased recycling has reduced the demand for primary materials produced by mines. Higher energy costs have led to a reduction in the physical amount of metals used per unit of manufactured products, especially in the transportation sector. Materials technology has brought forth a number of nonmetallic materials such as polymer composites and ceramics that are lighter and stronger than metals. These new materials are increasingly used as substitutes for metals in a variety of manufactures. In addition, new alloys combining metals such as titanium, tungsten, molybdenum and lithium with aluminum, steel and other traditional metals, provide strength and other desirable characteristics while economizing on metal content. Although the new metallic and nonmetallic materials have thus far been used only for high technology products, many of them, along with newer materials still in the development stage, are likely to find applications in products for mass consumption.

(2) The metals industries of the USA and other industrial countries have been losing control of metal prices in their national markets. Prices are no longer being set by producers that adjust production to declining demand, but are more and more determined by quotations on international metal exchanges, thereby putting the industrial country producers in direct competition with producers in developing countries that make no effort to adjust output to shifts in demand. This has occurred in copper, lead and zinc, and is beginning to take place in aluminum and nickel.

(3) Production for world markets has been shifting from the USA, Canada and Western Europe to countries with higher grade ores, lower energy costs and lower wages. Thus there is a shift in production of aluminum, copper, iron ore, lead and zinc to Australia, Brazil, Chile and the countries of the Asian Pacific Rim. In time, a larger share of the world copper market is likely to be supplied from Zambia and Zaire, which have the highest-grade copper resources in the world. Moreover, the developing countries producing mine products for shipment to industrial countries are integrating downstream to the production of aluminum, steel and a variety of fabricated and manufactured metal products.

(4) Low prices and low profitability of primary production are forcing American metal firms away from primary products and into downstream end-products and new metallic and nonmetallic materials. The traditional metal companies are finding it more profitable to import primary mine products and concentrate on producing upgraded materials.

(5) Although mine output in the USA is declining, it will continue to supply a significant portion of the US market for copper, iron ore,

lead and zinc. The US mines remaining in operation are reducing costs by modernizing and adopting new technology. There is also a trend in the world mining industry toward the development of small multi-deposit mines and smaller processing facilities. These developments will increase the flexibility of multi-mine firms in adjusting production to changing price and demand conditions.

(6) The role of multinational mining companies in Third World countries has declined over the past decade and their share of Third World output seems likely to continue to shrink. With a few exceptions, such as Chile, there are likely to be few large investments by multinationals in Third World countries. Foreign equity financing will tend to be avoided in favor of loan financing and joint ventures with SMEs. The development of new capacity by SMEs in Third World countries with serious debt problems may be constrained by a shortage of external capital since they will no longer be able to obtain government-guaranteed loans. This may make them more willing to seek multinational firms as partners in joint ventures.

(7) The level of world investment in mining is likely to remain low for several years until present overcapacity is eliminated and real prices rise sufficiently to cover full economic costs of production. This could mean temporary periods of high prices for some metals until capacity expands to meet the growth in demand. However, the long-term trend of real metal prices is not likely to be significantly upward in the foreseeable future. Higher prices will simply stimulate the development and application of new technology for economizing on the use of metals and substitution of nonmetallic materials. In addition, cost reductions through improved mining and processing technology, together with advances in exploration technology, will continue to hold down real costs and prices for minerals as they have for the past several decades.

Notes

1 In his analysis Trainer (1982, p. 51) uses much larger recoverable resources than those in Table 8.3.

Bibliography

Adams, F. Gerard and Klein, Sonia A. (eds) (1978), *Stabilizing World Commodity Markets* (Lexington, Mass.: Lexington Books).

American Bureau of Metal Statistics (1983), *Production and Consumption Data from Nonferrous Metal Data 1982* (New York).

Bureau of Mines (1985), *Mineral Facts and Problems, 1985* (Washington, DC: US Department of Interior).

Bureau of Mines (1986a), "Minerals in the World Economy," preprint from *Minerals Yearbook 1984* (Washington, DC: US Department of Interior).

Bureau of Mines (1986b), *Mineral Commodity Summaries, 1986* (Washington, DC: US Department of Interior).

Burke, W. Scott and Brokaw, Frank S. (1982), "Law at Sea," *Policy Review*, May, pp. 71–82.

Callot, F. G. (1981), "World Mineral Production and Consumption in 1978," *Resources Policy*, March, pp. 15–28. This article is an abridged version of a comprehensive report by F. G. Callot in the French journal, *Annales des Mines*, November/December 1980.

Carmen, John S. (1985), "The Contribution of Small-Scale Mining to World Mineral Production," *Natural Resources Forum*, vol. 9, no. 2, May, pp. 119–24.

Caves, Richard (1982), *Multinational Enterprise and Economic Analysis* (Cambridge: Cambridge University Press).

Chapman, Steven (1982), "Underwater Plunder," *New Republic*, April 20, pp. 17–20.

Charles River Associates (1977), *The Feasibility of Copper Price Stabilization Using a Buffer Stock and Supply Restrictions from 1953–1976* (Cambridge, Mass.: Charles River Associates).

Comptroller General (1979), *US Mining and Mineral Processing Industry: An Analysis of Trends and Implications* (Report to Congress) (Washington, DC: USGPO), October 31.

Congressional Budget Office (1985), *Environmental Regulation and Economic Efficiency* (Washington, DC: USGPO), March.

Congressional Research Service (1983), *Seventh Biennial Conference on National Materials Policy* (Report prepared for the House Committee on Science and Technology) (Washington, DC: USGPO), March.

Contreres, Raul (1984), "CODELCO's Projected Copper Supply, 1984–1990." *Quarterly Review* (Paris: CIPEC), April–June, pp. 47–57.

Crowson, Philip (1982), "Investment in Future Mineral Production," *CIPEC Quarterly Review* (Paris: Intergovernmental Council of Copper Exporting Countries), April–June, pp. 4–53.

Crowson, Philip (1983), "Aspects of Copper Supply: Past and Future," *CIPEC Quarterly Review* (Paris: Intergovernmental Council of Copper Exporting Countries), January–March, pp. 38–47.

Cunningham, Simon (1981), *The Copper Industry in Zambia: Foreign Mining Companies in a Developing Country* (New York: Praeger).

Dick, Rolf (1985), "Deep-Sea Mining Versus Land-Based Mining: A Cost Comparison," in J. B. Donges (ed.), *The Economics of Deep-Sea Mining* (Berlin: Springer Verlag).

Dunning, John H. (1981), *International Production and the Multinational Enterprise* (London: Allen & Unwin).

Economic Consulting Services (1981), *Nonfuel Mineral Policies of Six Industrialized Countries* (Washington, DC: US Department of Commerce).

Engineering and Mining Journal (1985), "Company-Government Disputes May Close Ok Tedi," March, pp. 27–8.

Escobar, Janet K. (1982), "Comparing State Enterprises Across International Boundaries: The Corporacion Venezolana de Guyana and the Companhia Valle Do Rio Doce," in Jones (1982).

Freeport-McMoRan, *Annual Report* (New York: Freeport-McMoRan), various issues.

Garnaut, Ross and Clunies-Ross, Anthony (1983), *Taxation of Mineral Rents* (Oxford: Clarendon Press).

Gillis, Malcolm, Jenkins, Glen P. and Lessard, Donald R. (1982), "Public Enterprise Finance: Toward a Synthesis," in Jones (1982).

Gladwin, Thomas A. (1985), *Environmental Aspects of the Activities of Transnational Corporations: A Survey* (New York: United Nations).

Gladwin, Thomas A. and Welles, John G. (1976), "Environmental Policy and Multinational Corporate Strategy," in Walter (1976).

International Finance Corporation, *Annual Report* (Washington, DC), various issues.

International Trade Commission (1984), *Unwrought Copper* (Report to the President on Investigation No. TA-201-52) (Washington, DC: USGPO), July.

Jones, Leroy P. (ed.) (1982), *Public Enterprise in Less Developed Countries* (Cambridge: Cambridge University Press).

Labys, W. (1982), "The Role of State Trading in Mineral Commodity Markets," in Marion Kostecki (ed.), *State Trading in International Markets* (New York: St. Martin's Press).

Larsen, David L. (1982), "The Reagan Administration and the Law of the Sea," *Ocean Development and International Law Journal*, vol. 11, no. 3/4, pp. 197–330.

Leaming, George F. (1983), paper presented at Second International Symposium on Small Mine Economics and Expansion, Helsinki; paper summarized in article entitled "Second Small Mine Symposium Held in Helsinki," *World Mining*, October.

Maizels, Alfred (1982), *Selected Issues in the Negotiation of International Commodity Agreements: An Economic Analysis* (Geneva: UNCTAD).

Malone, James L. (1982), "U.S. Participation in the Law of the Sea Conference," *Bulletin*, (Washington, DC: US Department of State), May.

Manners, Gerald (1981), "Our Planet's Resources," *The Geographical Journal*, vol. 147, part I, March.

McPherson, Charles P. and Palmer, Keith (1984), "New Approaches to Profit Sharing in Developing Countries," *Oil and Gas Journal*, June 25.

Mikesell, Raymond F. (1983), *Foreign Investment in Mining Projects: Case Studies of Recent Experiences* (Cambridge, Mass.: Oelgeschlager, Gunn and Hain for Fund for Multinational Management Education).

Mikesell, Raymond F. (1984), *Petroleum Company Operations and Agreements in the Developing Countries* (Washington, DC: Resources for the Future).

Mikesell, Raymond F. (1985), "Economic Stockpiles for Dealing with Vulnerability to Disruption of Foreign Supplies of Minerals," *Materials and Society*, vol. 9, no. 1, pp. 59–128.

Mikesell, Raymond F. (1986a), *Stockpiling Strategic Materials: An Evaluation of the National Program* (Washington, DC: American Enterprise Institute).

Mikesell, Raymond F. (1986b), "Materials Industries: Financing," in Bever, Michael D. (ed.), *Encyclopedia of Materials Science and Engineering*, Vol. 8 (Oxford: Pergamon Press), pp. 2844–9.

Mikesell, Raymond F. (forthcoming), "New Taxation Formulas in Mine Investments: Sharing the Risks and the Rents," paper given at conference sponsored by Institute for Foreign and International Trade Law, Frankfurt, Germany, June 18–21, 1986 (to be published).

Mikesell, Raymond F. (1987), *Nonfuel Minerals: Foreign Dependence and National Security* (Ann Arbor, Mich.: University of Michigan Press for the Twentieth Century Fund).

Ministry of Natural Resources (1980), *The Future of Nickel and the Law of the Sea*, Mineral Policy Background Paper no. 10 (Ontario: Canadian government).

Petrick Associates (1980), "Pueblo Viejo," in *The Economics of Minerals and Energy Projects* (Evergreen, Colo).

Pintz, William S. (1984), *Ok Tedi: Evaluation of a Third World Mining Project* (London: Mining Journal Books).

Radetzki, Marion (1983), *State Mineral Enterprises: An Investigation into their Impact on International Mineral Markets* (Washington, DC: Resources for the Future).

Richardson, Elliott L. (1982), "U.S. Foreign Policy and the Law of the Sea," House Committee on Foreign Affairs, *Hearings*, 97th Congress, 2nd session (Washington, DC: USGPO), June-September.

Richardson, Elliott L. (1983), "The United States Posture toward the Law of the Sea Convention: Awkward but not Irreparable," *San Diego Law Review*, vol. 20, no. 3, April, pp. 505–17.

Rosenkranz, R. D., Boyle, Jr., E. H., Porter, K. E. (1983), *Copper Availability–Market Economy Countries: A Minerals Availability Program Appraisal* (Washington, DC: US Department of Interior, information circular 8930).

Siedenburg, W. (1985), *Copper Quarterly* (New York: Smith Barney Harris Upham), July 10.

Southern Peru Copper Corporation (1980–4), *Annual Report* (New York).

Stobart, Christopher (1984), "The Effects of Government Involvement on the Economics of the Base Metals Industry," *Natural Resources Forum*, July, pp. 259–66.

Takeuchi, K., Strongman, J., Maeda, S. (1986), *The World Copper Industry: Its Changing Structure and Future Prospects*, Staff Commodity Working Paper no. 15 (Washington, DC: World Bank), November.

Trainer, F. E. (1982), "Potentially Recoverable Resources: How Recoverable?" *Resources Policy*, March, pp. 41–52.

UNCTAD Secretariat (1980), *The ITA in Operation during 1956–79* (Geneva: UNCTAD), April.

United Nations (1982), *United Nations Conference on the Law of the Sea*, Third UN Conference on the Law of the Sea (A/Conf.62/122) (New York: United Nations), October 7.

US Department of Commerce, *Survey of Current Business* (Washington, DC: US Department of Commerce), various issues.

US Department of State (1982), *Bulletin* (Washington, DC: US Department of State), March, pp. 54 ff.

Vernon, R. and Levy, B. (1982) "State-Owned Enterprise," in Leroy P. Jones (ed.), *Public Enterprise in Less Developed Countries* (Cambridge: Cambridge University Press).

Wall Street Journal, (1985a), "Alcoa Intends to Set Up New Venture in Further Effort to Diversify Its Lines," September 30.

Wall Street Journal (1985b), "Tin Price-Support Ills Shake Market," November 1.

Walter, Ingo (1975), *International Economics of Pollution* (London: Macmillan).

Walter, Ingo (ed.) (1976), *Studies in International Environmental Economics* (New York: Wiley).

Walter, Ingo (1984), "Project Finance: The Lender's Perspective," in D. W. Pearce, H. Siebert and I. Walter (eds), *Risk and the Political Economy of Resource Development* (New York: St Martin's Press).

White House (1985), *Press Release*, July 8.

Wilson, Forbes (1981), *The Conquest of Copper Mountain* (New York: Atheneum).

World Bank (1978–85), *Annual Report* (Washington, DC: World Bank).

World Bank (1983), *The Outlook for Primary Commodities*, Staff Commodity Working Paper no. 9 (Washington, DC: World Bank).

World Bank (1986), *Half-Yearly Revision of Commodity Price Forecasts and Quarterly Review of Commodity Markets for December 1985* (Washington, DC: World Bank), January.

Zuleta, Bernard (1983), "The Law of the Sea after Montego Bay," *San Diego Law Review*, vol. 20, no. 3, April, pp. 475–87.

Index